GEOCHEMICAL AND HYDROLOGIC PROCESSES AND THEIR PROTECTION

GEOCHEMICAL AND HYDROLOGIC PROCESSES AND THEIR PROTECTION

THE AGENDA FOR LONG-TERM RESEARCH AND DEVELOPMENT

Edited by
Sidney Draggan, Ph.D.,
John J. Cohrssen,
and Richard E. Morrison

New York
Westport, Connecticut
London

Library of Congress Cataloging-in-Publication Data

Geochemical and hydrologic processes and their protection.

Findings and recommendations of the Expert Panel Meeting on Geochemical and Hydrologic Processes and Their Protection, held May 1984 in Washington, D.C., presented to the Council on Environmental Quality, Interagency Subcabinet Committee on Long-term Environmental Research.

Includes index.

1. Geochemistry. 2. Hydrology. 3. Atmosphere. 4. Pollution. 5. Environmental protection. I. Draggan, Sidney. II. Cohrssen, John J. III. Morrison, Richard E. IV. Expert Panel Meeting on Geochemical and Hydrologic Processes and Their Protection (1984 : Washington, D.C.) V. Council on Environmental Quality (U.S.). Interagency Subcabinet Committee on Long-term Environmental Research.

QE15.G352 1987 551.9 87–6974

ISBN 0–275–92339–8 (alk. paper)

This material is based upon work supported by the Council on Environmental Quality and the National Science Foundation under the project "Council on Environmental Quality Conference on Long-Term Environmental Research and Development." The government has certain rights in this material.

Any opinions, findings, and conclusions or recommendations are those of the authors and do not necessarily reflect the views of the Council on Environmental Quality or the National Science Foundation.

Library of Congress Catalog Card Number: 87-6974
ISBN: 0–275–92339–8

First published in 1987

Praeger Publishers, 521 Fifth Avenue, New York, NY 10175
A division of Greenwood Press, Inc.

Printed in the United States of America

The paper used in this book complies with the Permanent Paper Standard issued by the National Information Standards Organization (Z39.48–1984).

10 9 8 7 6 5 4 3 2 1

CONTENTS

PREFACE

Over a period of fourteen months a large, diverse group of experts representing government agencies, industry, public interest groups, and the environmental science community worked on a wide range of tasks to produce this volume. Their overall task was to discern the current state of long-term environmental research and development (R&D) activities undertaken by government, industry, and universities, and to develop a set of guiding principles and priorities for this heretofore neglected component of environmental R&D.

On March 18, 1985, the findings and recommendations of four expert panel meetings and of a summary conference were presented to the Council on Environmental Quality (CEQ) Interagency Subcabinet Committee on Long-Term Environmental Research, established by CEQ Chairman A. Alan Hill in January 1985. This volume presents the findings and recommendations of the Expert Panel Meeting on Geochemical and Hydrologic Processes and Their Protection as well as the report prepared by the chairmen of the four expert panels.

Although published in four separate volumes, the findings and recommendations produced in this exercise are meant to be viewed as a whole. It is hoped that the readers of this volume will be led to inspect the outputs of the other expert panel meetings which have been produced as companion volumes.

ACKNOWLEDGMENT

This project to develop guiding principles and priorities for long-term environmental research and development (R&D) activities depended on the advice and participation of over 300 individuals intimately involved in long-term environmental R&D. Whether providing input on expert panel topics and participants or serving as panelists, background paper authors, or manuscript reviewers, these individuals helped to assure the success of the undertaking. The editors wish to express their thanks to these experts whose experience and cooperation have led to the four volumes in this series.

STATEMENT OF THE CHAIRMAN OF THE COUNCIL ON ENVIRONMENTAL QUALITY

This publication of the Council on Environmental Quality, *Report on Long-Term Environmental Research and Development,* is the culmination of an important process during which the collective wisdom of scientists in Federal agencies, private industry, academia, and environmental organizations was solicited, debated, synthesized, and summarized. Although the focus is on long-term environmental research, the observations and recommendations apply as well to shorter-term environmental research, since there are no distinguishable boundaries that uniformly differentiate long-term from other environmental research.

The Council on Environmental Quality routinely disseminates environmental research results in its annual report and by other means. The information in this Long-Term Environmental Research and Development Report is particularly important since science and technology provide the conceptual bases for action that can minimize future environmental risks. The research areas identified in the Report are, and I expect will continue to be, the focus of increased attention. It would be of great interest to the Council to learn how readers find this Report. Please let us know by writing to CEQ, 722 Jackson Place, NW, Washington, DC 20006.

A. Alan Hill

REPORT OF THE CHAIRMEN

EXECUTIVE SUMMARY

Scientists knowledgeable about human health and the quality of the natural environment recognize that man-made pressures placed upon natural resources, both living and nonliving, are more severe than previously suspected. Significant gaps in the existing scientific knowledge base cause or exacerbate acknowledged problems of environmental management. Long-term environmental and health research and development (R&D) is needed to resolve these scientific uncertainties, to establish baseline health and environmental parameters, to overcome lack of understanding of short-term variations in natural systems, and to identify long-term trends and relate them to their causes.

The Council on Environmental Quality (CEQ) convened a series of four two-day scientific and technical panel meetings to address specific clusters of long-term environmental R&D topics and to discern major scientific issues warranting enhanced government or private sector attention during the remainder of the century. This effort was requested by the Environmental Protection Agency (EPA) and was substantially assisted by the National Science Foundation (NSF). Additional financial assistance was provided by the Department of Energy, the National Institute of Environmental Health Sciences, and the Nuclear Regulatory Commission.

As panel chairmen, we caution that the long-term environmental and health R&D recommendations presented herein should not be viewed collectively as the comprehensive national R&D agenda for the remainder of the century; rather, they reflect those R&D issues to which substantial *increased* attention should be devoted. We recommend that the Federal government accord long-term environmental research heightened priority and a level of support sufficient that research findings will contribute to the attainment of broad National goals. This level of support need not require additional resources; it will require, however, a continuity in many existing research programs, a continuing commitment to scientific excellence, and an adequate institutional framework.

Principal long-term environmental R&D issues identified as warranting particular emphasis include the following:

- Improving the quality and cost-effectiveness of physical, chemical, and biological monitoring programs to test scientific hypotheses about how environmental systems operate and interact;
- The establishment of comprehensive "centers of excellence" at existing universities or other research-based institutions;
- Continued progress in the use of molecular epidemiology techniques to detect and measure the interaction of foreign chemicals with easily accessible normal human constituents;
- Expanded research on geohydrological processes data at the soil/water/air/hazardous waste interface;
- Identification of genetic and other factors that account for individual differences in susceptibility to environmental agents;
- Development of underlying scientific principles for evaluating the toxicity of mixtures of chemicals;
- Evaluation of consequences of human or ecological exposure to chemical substances used in, or organisms that are chemical byproducts of emerging technologies;
- Expanded collection of 10-year observations of the background physical, chemical, and ecological variations in fresh waters, oceans, and the atmosphere;
- Determination of intermedia pollutant transfer rates for description and prediction of the subsequent fate of toxic chemicals in the environment;
- Examination of the behavior and biological effects of chemicals in environmental media;
- Characterization of the role of biological and physical/chemical linkages and processes in specific biogeochemical cycles;
- Identification of the biological interactions and ecosystem processes most sensitive to global pollutants and environmental stresses;
- Development of biological inventories and baseline studies of ecosystem structures, functioning, and linkages; and
- Improvement of the scientific basis for quantitative risk assessments, and of procedures for extending them to human health effects other than cancer, such as systemic toxicity.

BACKGROUND

CEQ, at the request of the Environmental Protection Agency (EPA) and with assistance from the National Science Foundation (NSF), convened a series of four two-day scientific and technical panel meetings to address specific clusters of long-term environmental research and development (R&D) topics[1] and to discern major scientific issues warranting enhanced government or private sector attention during the remainder of the century. Overseeing this activity was an interagency subcabinet Committee on Long-Term Environmental Research (the Committee), chaired by CEQ Chairman A. Alan Hill. Support for the effort was provided by NSF, EPA, the Department of Energy, the National Institute of Environmental Health Sciences, and the Nuclear Regulatory Commission.

At each meeting, a panel of experts discussed a set of four substantively interrelated research topics and associated topic groupings, each of which was originated through consultation with Federal agencies. Panelists included public and private sector representatives, among whom was a rapporteur whose function was to provide a report summarizing the discussion. Background papers were prepared for each panel, and during the panel meetings the authors responded to questions.

After all panel meetings had taken place, a fifth meeting was convened at which the panel chairmen, rapporteurs, and CEQ and NSF staff met to summarize the conclusions and recommendations of the respective panels and to identify overriding or cross-cutting long-term environmental R&D issues. This draft report is the output of that meeting. CEQ expects, following Committee review of this draft and appendices, to disseminate a report on long-term environmental R&D.

When published in final form, the CEQ report will include the background papers for the four panel meetings, rapporteur reports summarizing the discussions of each of the four panels, and this report, together with any agency comments received on the circulated draft.

THE CONTEXT

Scientists knowledgeable about health and the environment recognize that man-made pressures placed upon natural resources, both living and nonliving, are

---

[1] As used here, long-term environmental R&D includes three kinds of activities: (1) anticipatory research, designed to identify potential environmental problems before they occur; (2) investigations of a continuing nature, such as ecological baseline studies or epidemiological studies, which may require a period of up to several decades to complete; and (3) fundamental research, the output of which may advance basic understanding of health- or environment-related processes.

more severe than previously suspected. There is mounting evidence, for instance, that potentially health-threatening groundwater contamination is a problem of increasing concern in the United States, and that toxic chemicals in hazardous waste dumps and underground storage may pose serious health and environmental threats and create public anxiety. Analogous problems with toxic chemicals have been experienced with air and water and in the workplace. During the past 40 years a wide variety of new synthetic chemicals have been introduced, some of which appear to pose serious acute and chronic health effects. Of similar concern is evidence of potential and perhaps irreversible damage to such important natural processes or properties as biogeochemical cycling or biologic diversity. Examples of possible changes in natural ambient levels that may be precursors of damage to health or ecosystems include the observed gradual increase in atmospheric carbon dioxide concentrations, suspected increases in atmospheric methane concentrations, possible increases in non-urban concentrations of ozone and carbon monoxide (current monitoring efforts detect primarily urban changes), observed increases in rates of species extinction in various parts of the world, and substantial ecosystem degradation in estuarine environments.

Ignorance about many scientific questions has resulted in acknowledged problems of environmental management, such as inappropriate regulation in the face of data uncertainties or heightened public anxieties. Long-term environmental and health research is needed to resolve scientific uncertainties, to establish baseline health and environmental parameters, to overcome lack of understanding of the short-term variations in natural systems, and to identify long-term trends and relate them to their causes.

For a variety of reasons, current incentives for private sector and governmental support of environmental and health R&D favor short-term approaches. Government agency research programs are generally designed to support mission goals of the agencies sponsoring them, resulting in relatively short-term research planning horizons that do not extend beyond immediate regulatory or programmatic requirements. Similarly, corporate research efforts frequently support near-term product development strategies and are necessarily reflective of annual (or shorter) profit and loss statements. Current government and public concern over such environmental problems as acid deposition phenomena (acid rain) illustrates the fallacy in continued reliance on short-term research design. Although the potential environmental and health effects of acid particulates were pointed out years ago, relatively little research attention was devoted to following up on early studies noting these effects, and commitment of resources could not be justified on the basis of then-current regulatory strategies. Accordingly, long-term acid rain research programs were deferred; had they been undertaken a decade ago they might by now have been yielding information and predictive models of use to current regulators and policymakers.

We believe there is need for a greater resource commitment to and better direction, coordination, and interdisciplinary integration of long-term environmental and health R&D. Improvements in environmental management will flow from better characterization of environmental phenomena, increased understanding of basic mechanisms, and the development of more meaningful measures of hazard or harm assessment. Our panelists noted the need for good long-term monitoring data and accompanying quality assurance to evaluate models used for understanding processes and environmental trends. They believe that lack of validated monitoring time series data, based upon even crude health and environmental measures, has impeded the expansion of fundamental research programs. Further, they believe, the piggybacking of monitoring onto ongoing research activities may well provide useful information at little marginal cost.

We believe that modeling can be an integrating force for the environmental research community, in that the imposition of modeling requirements yields helpful insights in identifying needs and opportunities for new research. However, the use of models must be accompanied by continuing efforts to validate them. For example, in constructing quantitative risk assessment (QRA) models, it becomes clear that the unknowns in the QRA can only be satisfied by continued basic research on biochemical mechanisms. Efforts to fill modeling gaps can, in turn, provide help in determining the need for and priority of data collection activities.

General acceptance of these observations led to a consensus that the current institutional framework and organization of government resources may not assure that needed long-term research projects can be started, continued, and completed with adequate funding on a multiyear basis. Only with research continuing over many years, and for projects that extend over a substantial period of time and that are focused on fundamental issues, will the Nation be able to develop the credible and necessary expertise to better rationalize environmental management policies. Environmental science needs a critical mass of talent and resources to effectively approach the challenge of understanding complex environmental and health phenomena. The need to integrate data gathering, modeling, and environmental impact assessment and QRA efforts may require the establishment of centers of excellence, at which this critical mass of talent and resources can be assembled.

Over the long term, environmental science may provide answers to such unresolved questions as the effects of subtle changes on ecosystems, the effects of new technologies such as biotechnology and microelectronics on human health, and the effects of global contaminants on ecological processes.

CHAIRMEN'S RECOMMENDATIONS

Monitoring

Monitoring yields essential current and time series information on the status of environmental systems, information used both for environmental management and for regulatory compliance. Information gained from monitoring forms the basis for the testing of scientific hypotheses about how environmental systems operate and interact. Panelists believe that the lack of coordinated scientific resources providing institutional sophistication in environmental monitoring, modeling, biostatistics, and ecostatistics account for such problems as nonstandardization of monitoring practices and failure to monitor for parameters of greatest importance or relevance.

In the absence of validated data, modeling is frequently the only source of guidance to environmental managers for describing or predicting complex environmental events. Invariably one or more elements of necessary environmental data are missing, either because we lack the understanding or tools to measure the phenomena or because the cost of measurement is prohibitive. To the extent that monitoring data necessary for model-building are not validated or not reliable, models lack predictive value. Environmental models also frequently suffer from over complexity or from lack of standardization.

In spite of significant efforts expended on the collection of environmental data by a variety of unrelated and uncoordinated state and Federal agencies, no adequate system exists for the centralized collection, storage, maintenance, and quality control of such data. We recommend that CEQ foster an evaluation by an appropriate organization of existing physical, chemical, and biological monitoring programs (and extant data associated with them) to identify and stimulate research and development on improving the quality and cost-effectiveness of monitoring programs. Particular emphasis should be placed on determining requirements for biological and environmental monitoring, on identifying pollutants such as toxic chemicals for sampling, and on determining information and statistical requirements for environmental models.

Institutional Capability

There is an evident disparity between the current mix of institutional arrangements and resource availability to design, conduct, and manage long-term environmental research, on the one hand, and the long-term research needs identified herein, on the other. The establishment of well-funded "centers of excellence" at existing universities, such as those now supported by the National Institutes of Health and by EPA, was discussed as a possible means to redress this imbalance. The National Institute of Environmental Health Sciences represents one example of a governmental institution that has successfully supported integrated research over a long term. Another example is the Long-Term Ecological Research Program supported by NSF's Division of Biotic Systems and Resources. Laboratories owned and operated by EPA, where the research staff are government employees, provide another model. Research laboratories operated by the Federal government on contract to universities, university-based consortia, or industry-based research organizations—such as the National Center for Atmospheric Research, funded by NSF, or the national laboratories supported by the

Department of Energy—provide still a third model, one at which government-sponsored research is funded on a multiyear basis.

To supplement government support, we recommend that attention be given to seeking sources of environmental and health research funding and ideas from the private sector to the extent possible, and that effort be given to coordinating government-sponsored research with research performed or supported by industry-specific organizations such as the Electric Power Research Institute, the Health Effects Institute, the Chemical Industry Institute of Toxicology, and the Gas Research Institute.

Molecular Epidemiology and Exposure Estimation

Molecular epidemiology is based upon measurement in the exposed individual of the interaction of a toxic chemical (or its derivative) with a tissue constituent or a tissue alteration resulting from exposure to the chemical. It thus provides an indirect measure of individual exposure.

There has been significant recent progress in refining means for detecting and measuring the interaction of foreign chemicals with easily accessible normal human constituents, such as chemical carcinogen interactions with deoxyribonucleic acid (DNA). It is now sometimes possible to detect a few altered DNA molecules out of the millions in a cell. Along with other developments for detection of altered chromosomes in human cells, these refinements mark important advances in our ability to detect and quantitate human exposure to foreign chemicals. Potential benefits from molecular epidemiology are so great as to justify an expansion of the national molecular epidemiology research effort. This research may eventually yield accurate markers of exposure, and information on the relationship between exposure and adverse health outcomes.

Hazardous Waste Sites

Panelists indicate that substantial geohydrological processes data gaps exist at the soil/water/air/hazardous waste interface, and that very little is known or fully understood about underlying processes. Current hazardous waste site cleanup actions are routinely commenced in the absence of knowledge about how chemicals move through the soil, how they are transformed, and how modeling can successfully be used to predict air and water contamination. We recommend that government agencies, particularly EPA and DoD, integrate well-defined research studies into plans for measurement and remedial action programs at selected hazardous waste sites.

Genetic Diversity/Susceptibility and Biological Mechanisms

Susceptibility to chemical toxicants varies widely in human host populations. Such host factors as genetic diversity, current or prior disease, sex, and age contribute to this variation. There are approximately 2,000 genetically identifiable human diseases, and there are firm theoretical foundations for the hypothesis that genetic conditions are likely to enhance (sometimes dramatically) the risk to affected individuals of developing environmentally or occupationally associated adverse health effects; however, the extent of risk enhancement is currently unknown.

It is recognized that identification of different susceptibilities as a result of genetic differences may generate or exacerbate social problems. Although this variation creates difficulties for the research establishment, it should not be allowed to inhibit scientific inquiry. Many of the differences in susceptibility derive from major differences in the way individuals alter (intensify or decrease) chemical toxicity through their different enzymes (active body constituents that alter foreign chemicals). We recommend that a research effort be mounted to identify genetic and other factors important for individual differences in susceptibility to environmental agents, to assess the health impacts of such differences, and to examine the means (such as differences in enzymatic activity) by which these differences become manifest.

Mixtures

Exposures of humans, other animals, and the natural environment to pollutants rarely involve single substances. Human and nonhuman organisms are generally exposed to mixtures of chemicals, whether from waste dumps, contaminated water, or ambient or workplace air. When single agents can be identified as of predominant importance, the problem is simplified. Often, however, this is not the case. Regrettably, there are no valid general rules for determining the presence of synergism in inhibitory action, and any study requiring a mixture by mixture analysis would be impossibly complex. We recommend, therefore, a major effort to define underlying general scientific principles for evaluating the toxicity of mixtures of chemicals.

Anticipating the Impacts of Emerging Technologies

Two emerging technologies—biotechnology and microelectronics—were examined. The first deals with the new field of genetic engineering, which promises to yield more efficient or new medicinal products, agricultural products, chemicals, and other products. The second involves the production of microelectronic chips with the use of a variety of virtually unstudied (exotic) chemicals, such as gallium arsenide, silicon, and halogenated hydrocarbon solvents. We recognize that there may be health and environmental consequences associated with emerging technologies, and we therefore recommend that these possible consequences be evaluated.

Fundamental Research in Fresh Water, Ocean, and Atmospheric Cycles

We lack 10-year observations of the background physical, chemical, and ecological variations in fresh waters, oceans, and the atmosphere that are adequate to distinguish natural changes from perturbations attributable to human activities. We lack continental-scale background information on historic chemicals, or even on currently important chemicals in the environmental, particularly coastal waters and the atmosphere. We are also ignorant of the normal range of variations through which ecosystems and their components pass. In addition to obtaining better information on background concentrations or biological variation, we also need to examine the processes that account for ecological variations and how pollutants interfere with those processes.

We recommend expanded collection of 10-year observations of these phenomena to better enable us to distinguish natural fluctuations from those caused by anthropogenic activities, at study sites selected to permit simultaneous observation of contaminated and close-by or ecologically similar uncontaminated areas. For study of coastal waters, several estuaries have been identified as receiving or having received significant amounts of pollutants; the fates of these pollutants should be continuously monitored. We recommend that long-term continental-scale and global-scale baseline atmospheric measurements be gathered for polluting gases such as ozone, nitrous oxides, sulfur dioxide, and hydrocarbons, and for particulates. These pollutants should be studied to determine their effects on climate, visibility, and variables that affect human health and the quality of life. We also recommend an expanded research effort to more effectively describe the transport and transformation of toxic organic compounds with respect to their physical, chemical, and biological effects in the subsurface soil and groundwater environments. Subsurface and groundwater research activities should be focused on better understanding of the interactions of groundwater with the solid phases of rocks and soils; on developing a quantitative understanding of the variation of the sorption and vapor pressure of organic compounds with soil moisture, particularly below 10 percent soil water content; and on the fundamentals of hydraulics and geochemistry for low-permeability materials. Given the expense of groundwater monitoring, research could most efficiently be coordinated with existing compliance monitoring activities.

Intermedia Transfer

Pollutants move from one receiving medium to another, and their concentrations in any given reservoir depend on the rates of these transfers. Predictions of pollutant distribution within the environment are based on quantitative assessment of these movements. At present, quantitative rate measurements of these intermedia transfers are still rudimentary.

We recommend studies to determine intermedia transfer rates for both description and prediction of the subsequent fate of toxic chemicals in the environment. Of particular interest are aerosol formation rates of chemically reactive organic pollutants and the effects of surface active agents on the air/water interface, precipitation scavenging of aerosols in the field, dry deposition as a function of topological (surface) roughness, volatilization and adsorption as a function of temperature and moisture content, and the deposition and resuspension of sediments. Determining the chemical composition of aerosol particles and the size distribution of suspended solids in the liquid phase will also help in understanding these transfer rates. We further recommend that the results of intermedia monitoring and modeling studies be used to develop screening procedures to determine under what environmental conditions and with which chemical substance characteristics it is necessary to consider intermedia transfers. We also recommend linking intermedia transport processes to the biological effects observed in receptor organisms affected by exposure to the transferred chemicals; monitored receptor organisms should include humans, commercially important species, and whole ecosystems. Finally, we recommend that anticipatory research on the intermedia nature of pollutant transfer be used in the consideration of the respective cost efficiencies of alternative multimedia control strategies. For instance, the Toxic Substances Control Act, with proposed or hypothesized amendments, should be evaluated as a basis for regulating toxic and hazardous wastes by means of multimedia control strategies.

Assimilative Capacity

In the United States billions of tons of solid waste (of which a small percentage can be designated as hazardous) are released to the land, air, surface waters, and groundwaters. Wastes also move from one medium to another and ultimately reside or degrade in various media. Assessments must be made of the cross media risks associated with waste treatment and disposal practices. Each environmental medium, and the biological communities that reside within it, is assumed to have a site-specific set of responses to different levels of waste loadings. Various wastes differ on many characteristics including toxicity, mobility, degradation rates, and their ability to be recycled or otherwise treated. Presumably some amount of certain wastes may be assimilated by the different media without adverse impacts on human health and environmental quality. Identification of the lower limits at which public health, ecosystem integrity, or the most sensitive species populations are affected is crucial to a determination of environmental quality.

Although we recognize that considerable research has been done in some of these subject areas, we recommend that research be undertaken to determine the behavior and biological effects of chemicals in environmental media as a basis for setting levels of release for protection of human health and environmental quality. In addition, the determination of these levels should include a consideration of such changes in the physical environment as global warming, decreases in atmospheric visibility, and their impact during extremes of temperature and moisture encountered after one- to five-decade levels of variations.

Global Biogeochemical Cycles

Global biogeochemical cycles are essential to the maintenance of the biosphere. Continuous interaction between biotic and abiotic components of the Earth distinguishes it from other planets and, thus, the term biogeochemical cycling. We recommend the conduct of long-term fundamental studies to characterize the role of biological and physical/chemical linkages and processes in specific biogeochemical cycles (for example, in carbon, nitrogen, phosphorus, sulfur, and oxygen cycles), as well as in their interactions with each other and with climate to build on the results of previous research.

Global Pollutants and Impacts on Ecological Processes

Significant gaps in the environmental sciences knowledge base reflect our ignorance of cause/effect relationships between global pollutants and ecological processes. We recommend the conduct of long-term studies to identify the biological interactions and ecosystem processes that are most sensitive to specific past, present, and anticipated global pollutants and environmental stresses. The outputs of such studies might be helpful in clarifying the impact of pollutants on the total biosphere, and on global chemical processes.

Fundamentals of Ecosystem Structures and Processes

In order to identify those ecological phenomena that are related to the sustainability of ecosystem resources, and especially to the conservation of biological diversity, there is a need for long-term biological inventories and baseline studies of ecosystem structures and functioning. Especially needed are studies of processes that link different ecosystem types (for example, suburban/urban, agricultural/aquatic, industrial/agricultural linkages). We recommend the identification of ecosystem sites that can be studied and the establishment and maintenance of a national network of representative ecosystem sites, both those that are relatively undisturbed and those that are intensively managed.

Quantitative Risk Assessment

Quantitative risk assessment (QRA) has come to play a major role in regulatory processes and in risk management generally. Current QRA techniques are generally acknowledged to be crude and to lack the precision commensurate with the importance attached to the regulatory decisions frequently based on them, decisions that may be of great social and economic consequence.

Scientists now believe that improvements in QRA techniques are more likely to come from better understanding of biological processes than from improvements in mathematics. Far too little is now known, for instance, about basic pharmacodynamic and environmental mechanisms of toxic effects, about virtually any health (other than cancer) or ecological effects, and about actual exposure patterns; and about the potential of short-term biological screening to provide early prediction of effects. Further studies are needed on the independent validation of risk assessments, better understanding of decision processes, identification of scientific principles used in selection of assumptions and criteria, determination of the criteria used as a basis for risk assessment priorities, and improving the understanding of risks. We recommend that particular research attention be given to improving the scientific basis for risk assessments, and to procedures for valid extension to human health effects other than cancer, such as systemic toxicity.

PANEL TOPICS AND PANELISTS

HUMAN HEALTH IMPACTS AND THEIR MITIGATION (Published as ENVIRONMENTAL IMPACTS ON HUMAN HEALTH)

- Human Physiologic/Genetic Diversity
- Molecular Epidemiology
- Monitoring of Effects of Exposure to Health Hazards
- Human Health Impacts of Current and Emerging Technologies

GEOCHEMICAL AND HYDROLOGIC PROCESSES AND THEIR PROTECTION

- Surface Water/Groundwater Processes and Pollution
- Land/Soil Processes and Pollution
- Atmospheric/Oceanic Processes and Pollution
- Multimedia Toxic Substance/Hazardous Waste Research

ENVIRONMENTAL IMPACTS AND THEIR MITIGATION (Published as PRESERVING ECOLOGICAL SYSTEMS)

- Global/Biosphere Impacts
- Local Ecological Impacts
- Ecological Diversity
- Environmental Impacts of Current and Emerging Technologies

MONITORING, ASSESSMENT, AND ENVIRONMENTAL MANAGEMENT

- Data Generation, Collection, Analysis, and Interpretation
- Risk/Impact Assessment Techniques
- Modeling and Forecasting Techniques
- Environmental Management Approaches

HUMAN HEALTH IMPACTS AND THEIR MITIGATION

Human Physiologic/Genetic Diversity

"Priority Needs in the Development of Genetic Epidemiology"
Author: Dr. Mark Skolnick, University of Utah Medical Center

"Effects of Human Physiologic and Genetic Variability on the Development and Expression of Pollutant-Related Diseases"
Author: Dr. Elliot S. Vesell, Pennsylvania State University

Molecular Epidemiology

Molecular Epidemiology: Novel Approach to the Investigation of Pollutant-Related Disease"
Author: Dr. Frederica P. Perera, Columbia University

"The Value of Molecular Epidemiology in Quantitative Health Risk Assessment"
Author: Dr. Dale Hattis, Massachusetts Institute of Technology

Monitoring of Effects of Exposure to Health Hazards

"Problems in Demonstrating Disease Causation Following Multiple Exposure to Toxic or Hazardous Substances"
Author: Dr. Robert Dixon, U.S. Environmental Protection Agency

"The Value of Health Registries in Monitoring Pollutant-Related Disease"
Author: Dr. Paul Schulte, National Institute for Occupational Safety and Health

Human Health Impacts of Current and Emerging Technologies

"Improved Methods for Discerning Health Impacts of Current Technologies"
Author: Dr. James M. Robins, Harvard University

"Anticipating the Potential Health Impacts of Exotic Chemicals Associated with Emerging Technologies"
Author: Dr. Daniel T. Teitelbaum, Denver Clinic Medical Centers

"Identification and Control of Human Health Hazards Associated with Current and Emerging Biotechnology"
Author: Dr. Daniel Lieberman, Massachusetts Institute of Technology

Chairman

Dr. Norton Nelson	New York University Medical Center

Rapporteur

Dr. Edward J. Calabrese	University of Massachusetts

Panelists

Dr. Manning Feinleib	National Center for Health Statistics
Dr. Perry J. Gehring	Dow Chemical Company
Dr. Bernard Goldstein	U.S. Environmental Protection Agency
Dr. Irving Johnson	Eli Lilly and Company
Dr. James L. Liverman	Litton Bionetics
Dr. Brian MacMahon	Harvard School of Public Health
Dr. Mortimer L. Mendelsohn	Lawrence Livermore Laboratory
Dr. David P. Rall	National Institute of Environmental Health Services
Dr. Ellen K. Silbergeld	The Environmental Defense Fund

GEOCHEMICAL AND HYDROLOGIC PROCESSES AND THEIR PROTECTION

Surface Water/Groundwater Processes and Pollution

"Improving the Knowledge Base on Surface Water and Groundwater Resources"
Author: Dr. Wayne A. Pettyjohn, Oklahoma State University

"Protection of Surface Water and Groundwater Resources"
Author: Dr. Nathan Buras, University of Arizona

Land/Soil Processes and Pollution

"Identifying Knowledge Gaps on Land/Soil Processes: Hazardous Substances and the Land/Soil Resource"
Author: Dr. Louis J. Thibodeaux, Louisiana State University

"Advancing Knowledge on Protection of the Land/Soil Resource: Assimilative Capacity for Pollutants"
Author: Dr. Raymond C. Loehr, University of Texas

Atmospheric/Oceanic Processes and Pollution

"Global Air Chemistry and Continental-Scale Air Pollution: An Assessment of Long-Term Research Needs"
Author: Dr. Steven Wofsy, Harvard University

"Pollutant Impacts on Coastal Ecosystems"
Author: Dr. John M. Teal, Woods Hole Oceanographic Institution

Multimedia Toxic Substance/Hazardous Waste Research

"Scientific, Legislative, and Administrative Constraints to Multimedia Control of Toxic Substances and Hazardous Wastes"
Author: Leslie Sue Ritts, Esq., Morgan, Lewis and Bockius

"Modeling of Pollutant Transport and Accumulation in a Multimedia Environment"
Author: Dr. Yoram Cohen, University of California

Chairman

Dr. Edward D. Goldberg — Scripps Institution of Oceanography

Rapporteur

Dr. Michael S. Connor	U.S. Environmental Protection Agency

Panelists

Dr. Keros Cartwright	Illinois Geological Survey
Professor Henry P. Caulfield, Jr.	Colorado State University
Dr. Michael A. Champ	U.S. Environmental Protection Agency
Dr. Robert A. Duce	University of Rhode Island
Dr. Bruce A. Egan	Environmental Research and Technology, Inc.
Ms. Frances H. Irwin	The Conservation Foundation
Dr. Charles L. Osterberg	U.S. Department of Energy
Dr. Frederick G. Pohland	Georgia Institute of Technology

ENVIRONMENTAL IMPACTS AND THEIR MITIGATION

Global/Biosphere Impacts

"Maintaining the Integrity of Global Cycles: Requirements for Long-Term Research"
Author: Dr. William R. Emanuel, Oak Ridge National Laboratory

"On Toxins and Toxic Effects: Guarding Life in a Small World"
Author: Dr. George M. Woodwell, The Ecosystems Center

Local Ecological Impacts

"The Role of Basic Ecological Knowledge in Environmental Assessment"
Author: Dr. Stephen G. Hildebrand, Oak Ridge National Laboratory

"The Role of Basic Ecological Knowledge in the Mitigation of Impacts from Complex Technological Systems: Agricultural, Transportation, and Urban"
Author: Dr. Orie L. Loucks, Holcomb Research Institute

Biological Diversity

"Habitat Diversity and Genetic Variability: Are They Necessary Ecosystem Properties?"
Author: Dr. Elliott A. Norse, The Ecological Society of America

"Indicators of Change in Natural and Human-Impacted Ecosystems: Status"
Author: Dr. Frieda B. Taub, University of Washington

Environmental Impacts of Current and Emerging Technologies

"Improved Methods for Mitigating the Environmental Impacts of Current Technologies"
Author: Dr. Nicholas Clesceri, Rensselaer Polytechnic Institute

"Anticipating the Potential Environmental Impacts of Emerging Technologies: The Case of Genetic Engineering"
Author: Dr. Martin Alexander, Cornell University

Chairman

Dr. John E. Cantlon	Michigan State University

Rapporteur

Dr. Lev R. Ginzburg	State University of New York

Panelists

Dr. John W. Firor	National Center for Atmospheric Research
Dr. Marvin N. Glaser	Exxon Research and Engineering Company
Dr. Robert C. Harris	National Aeronautics and Space Administration
Dr. Dean L. Haynes	Michigan State University
Dr. Francis C. McMichael	Carnegie-Mellon University
Dr. Joseph F. Malina, Jr.	University of Texas
Dr. John M. Neuhold	Utah State University
Dr. J. C. Randolph	Indiana University

MONITORING, ASSESSMENT, AND ENVIRONMENTAL MANAGEMENT

Data Generation, Collection, Analysis, and Interpretation

''Development of Environmental Data Bases and Inventories''
Author: Dr. Don W. Hayne, North Carolina State University

''Improving Methods of Data Analysis and Interpretation for Environmental Management Programs''
Author: Dr. Richard E. Sparks, Illinois Natural History Survey

Risk/Impact Assessment Techniques

''Improving Quantitative Health Risk Assessment Techniques''
Author: Dr. Lester B. Lave, Carnegie-Mellon University

''Improving Qualitative and Quantitative Environmental Assessment Techniques to Support Environmental Management''
Author: Dr. Peter E. Black, State University of New York at Syracuse

Modeling and Forecasting Techniques

''Long-Term Research in Ecological Models for Environmental Management''
Author: Dr. Gordon L. Swartzman, University of Washington

''The Role of Forecasting Methodologies in Approaches to Environmental Management''
Author: Dr. John F. Ficke, The IT Corporation

Environmental Management Approaches

''The Role of Interdisciplinary Environmental Research in Natural Resource Management''
Author: Dr. David Pimental, Cornell University

''Innovative Environmental Management Approaches for Cost-Effective Control of Industrial Residuals''
Author: Richard A. Ferguson, Esq., The Skylonda Group, Inc.

Chairman

Dr. Richard M. Dowd	R. M. Dowd & Co.

Rapporteur

Dr. Richard N. L. Andrews	University of North Carolina

Panelists

Dr. A. Karim Ahmed	Natural Resources Defense Council
Dr. Bruce N. Bastian	Shell Oil Company
Dr. Edward J. Burger	Georgetown University Medical Center
Dr. James W. Gillett	Cornell University
Dr. Raphael G. Kasper	National Research Council
Dr. Franklin E. Mirer	United Auto Workers
Dr. Milton Russell	U.S. Environmental Protection Agency
Dr. Glenn W. Suter	Oak Ridge National Laboratory
Dr. Jaroslav J. Vostal	General Motors Research Laboratories

GEOCHEMICAL AND HYDROLOGIC PROCESSES AND THEIR PROTECTION

SUMMARY REPORT OF THE EXPERT PANEL MEETING ON GEOCHEMICAL AND HYDROLOGIC PROCESSES AND THEIR PROTECTION

INTRODUCTION

The meeting on Geochemical and Hydrologic Processes and Their Protection was included in the series of expert panel meetings because the observed changes in the environmental media of air, soil, and water are key indicators of the existence and ultimate effects of hazards to environmental quality and human health. Many connections with the long-term environmental research and development (R&D) topics addressed in the other panel meetings were evident. The Panel was charged with addressing long-term environmental R&D needs in four topic areas:

- Surface Water and Groundwater Processes and Pollution;
- Land/Soil Processes and Pollution;
- Atmospheric/Oceanic Processes and Pollution; and
- Multimedia Toxic Substance/Hazardous Waste Research.

THE CONTEXT

Knowledge of geochemical, atmospheric, and hydrologic processes enables us to understand how chemical substances we have released to the environment, or that we encounter from natural sources, are transported or transformed and thus inadvertently become part of the air we breathe, the food we eat, or the water we drink. These processes also affect the survival of plants, animals, and humans living at sites distant from where the substances were initially discharged, or initially existed. Environmental risk assessment requires an evaluation of exposure to chemical substances and of the nature and magnitude of effects associated with that exposure. Both categories of knowledge are essential to our capability to predict the consequences of the release of chemical substances to the environment. In order to predict exposure to chemical substances, environmental managers must determine quantitatively where these chemicals are discharged and how they are transported through and transformed by the environment. Long-term research is needed to define the relationships between geochemical, atmospheric, and hydrologic processes and the biological receptors of the transported chemical substances. We cannot preserve the health of biological communities or humans without knowledge of these transport and transformation processes.

Each year billions of tons of solid waste material, of which a small percent can be designated as "hazardous," are produced in the United States. Whether via landfills, surface impoundments, direct surface applications to land, incineration, fugitive emissions, or direct or indirect release to air, surface waters and groundwaters, these wastes are discharged to and ultimately reside or degrade in the three environmental media. Many wastes degrade rapidly. Some, however, persist or are transformed into substances that can cause damage.

Waste management strategies have assumed that soils can be used for interment of toxic and hazardous substances. But we still lack rudimentary knowledge of the quantitative importance of fundamental soil processes. We need to make an investment now in increasing our knowledge of these processes to prevent future waste disposal problems, problems that are serious in consequence. Wastes discharged improperly can increase our daily exposure to chemical substances that may be acutely or chronically toxic, or these substances may cause long-term environmental change to living or nonliving resources. Furthermore, these are long-term problems. Even with waste reduction processes, we will continue to generate some amount of residuals that require disposal.

Each medium has a specific capacity to assimilate some amount of the wastes it receives without unacceptable adverse impacts to human health or environmental quality. There are numerous examples, however, in which each of the environmental media has been overloaded in certain areas. In addition, there are transfers between the environmental media that may cause unanticipated environmental degradation. In some cases, the transfer from one medium to another has been the measure of concern; an example would be the movement of material from the soil to groundwater. This and other examples are described below.

Soil/Groundwater Manifestations

Large quantities of wastes are discharged to the soil from homes, cities, and industries. Researchers surveying hazardous wastes managed in 1981 under the authority of the Resource Conservation and Recovery Act estimated that of the total hazardous wastes disposed, 3 million wet tons were placed in landfills, 19 million tons were released to surface impoundments, and 32 million tons were discharged into deep injection wells. In addition, several million tons of hazardous wastes removed from hazardous waste sites over the next decade will need to be treated. Other less-visible threats to our soil/water resources include approximately 1.9 trillion gallons annually from septic tanks, 4.8 million dry tons/year of municipal sludge, and an unquantified amount of leakage from underground chemical storage tanks, particularly from the large number of gasoline storage tanks dotting our landscape. Although we have some information regarding the thousands of drinking water wells that have been forced to close, we have very little information concerning the magnitude of risks facing our groundwater supplies, both in terms of information on the status of whole aquifers and comprehensive data on concentrations of all organic compounds of concern.

Atmospheric Manifestations

The behavior of the atmosphere itself is affected by trace contaminants. Among these are methane, nitrogen oxides, and ozone. The atmosphere receives emissions from a number of different sources while providing rapid transport and dilution. Emissions have been increasing at a rapid rate from the surface of the Earth such that each year 10^8 tons of methane and 10^7 tons of nitrogen oxides are released to the atmosphere, a large percentage attributable to human activities. The emissions to the atmosphere of nitrous oxide and methane exceed known removal processes, and atmospheric concentrations of these pollutants are expected to increase during the next 10–50 years. Analysts examining data from polar ice cores have indicated that methane concentrations have doubled in the last 400 years. We are just beginning to understand the effects of these emissions, but it is expected that increased concentrations of nitrous oxide, methane, and other radiatively active gases will contribute to the warming of the Earth's climate. Ozone concentrations in rural U.S. locations are also thought to have increased, doubling since the 1950s, and episodes of high concentrations that can blanket the eastern United States are sufficient to cause damage to agricultural crops in laboratory experiments.

Atmospheric manifestations have terrestrial consequences as well. Ozone, nitrogen oxides, and sulfate deposition have all been hypothesized to be responsible for the decline of forests in Central Europe and the eastern United

States. Environmental exposure to airborne lead, benzene, and trichloroethylene, as well as to ozone and other substances, may pose significant health risks.

Coastal Ocean Manifestations

The open ocean can provide an enormous amount of dilution to minimize the risks presented by waste disposal, particularly those wastes dominated by high concentrations of metals or nutrients. Coastal estuaries, however, act as containment basins, collecting sediments and contaminants from rivers, direct discharges from coastal municipalities and industries, and atmospheric deposition. Several billion gallons of sewage effluent are discharged to our coastal waters each day, and millions of tons of solid wastes are dumped into the ocean each year.

The potential for estuaries to retain wastes can be seen in those estuaries that have become hazardous waste sites such as Commencement Bay, WA, and New Bedford Harbor, MA. In other estuaries, contamination of fish with kepone, PCBs, or dioxins has forced the closure of commercial and recreational fisheries, and bacterial contamination forces the closure of about one-quarter of the Nation's shellfish beds each year. Eutrophication in the Chesapeake Bay has been blamed for the decline of sea grass communities that serve as nursery areas for fish and help protect the coastline from erosion.

General Principles of Geochemical and Hydrologic Processes Research

The Panel's major findings fell into four general categories:

- The importance of an integrated approach to environmental problems;
- The need for long-term environmental research;
- The necessity of developing predictive models for determining the stresses placed on geochemical and hydrologic processes by different waste loadings; and
- The importance of determining the efficacy of different institutional strategies for environmental management.

Integrated Approaches

The Panelists endorse an integrated approach to the environmental problem of waste management, and feel that the approach must be

- integrated across the environmental media of air, soil, and water, and subcompartments within those media;
- integrated across substances present in the environment, both in how interactions between chemicals may affect the way these substances are

transported and transformed in the different media, and in how chemical interactions influence their effects on health and environmental quality; and
- integrated across disciplines addressing these problems, including disciplines in the natural sciences and social sciences concerned with the natural and institutional processes for controlling risks associated with alternative environmental management strategies.

The importance of the transfer of contaminants between environmental media has been documented repeatedly: volatilization of chlorinated organics from hazardous waste sites, sewage treatment plants, or contaminated shallow aquifers; the transfer of substances from air to soils and surface waters through precipitation and dry deposition; and the removal of contaminants from surface waters and groundwaters by sorption to soils or sediments and their subsequent rerelease during resuspension and desorption.

Although the results of intermedia transfer studies have often been qualitatively demonstrated, quantitative rate measurements of these processes are still rudimentary. Transfer rates of particular interest are aerosol formation of chemically reactive organic pollutants and the effect of surface active agents on the air/water interface, on precipitation scavenging of aerosols in the field, on dry deposition as a function of surface roughness, on volatilization and adsorption rates as a function of temperature and moisture content, and on the rates of deposition and resuspension of sediments. Determining the chemical composition of aerosol particles of different sizes and the size distribution of suspended solids in the liquid phase will also help in understanding these transfer rates.

The other CEQ Panels have recommended the study of long-term effects attributable to chemical interactions in complex mixtures of effluents. We endorse those recommendations. From preliminary results of ongoing studies, analysts believe that the movement of high concentrations of complex mixtures of wastes through soil differs markedly from laboratory models, in which the movement of dilute concentrations of single compounds is measured. Leaking containers at hazardous waste sites or accidental spills are prime examples of the occurrence of such high concentrations of complex chemical mixtures.

The study of geochemical and hydrologic process across different media obviously requires knowledge from a broad range of disciplines from soil science to atmospheric chemistry. The Panel also recognizes that natural science research of geochemical and hydrologic processes must be supplemented by social science research to facilitate interpretation of natural science results into laws and regulations that can be implemented and enforced.

Long-Term Environmental Research

In addition to the need for greater emphasis on integrated approaches, there are long-term needs for the study of transport and transfer processes and their biological consequences within each medium. We use the phrase "long-term environmental research" in all three definitional senses considered by CEQ: research that is fundamental in nature, that provides a continuing baseline, or that anticipates future research needs.

Fundamental Processes

The Panel finds that long-term research of fundamental processes is particularly necessary in the subsurface soil/groundwater environment. Our knowledge base is least advanced for this medium and currently inadequate to protect public health and the environment in all cases. Specific research is needed to better describe the transport and transformation of organic compounds considering hydraulic, biotic, and abiotic aspects in both homogenous and heterogenous soil matrices. For instance, water plays a dominant role in the soil environment, but we do not have a quantitative understanding of the variation of the sorption and vapor pressure of organic compounds with soil moisture, particularly below 10 percent soil water content. In addition, we know very little about the fundamentals of hydraulics and geochemistry for low-permeability materials, yet low-permeability sites are selected as optimal locations for the disposal of hazardous wastes. In addition, we need more information about the hydrologic properties of crystalline rocks and the importance of networks of fractures because most deep injection wells are located in these types of formations.

Continuing-Baseline Studies

The Panel finds that the short-term intensive studies at waste disposal sites can be linked with some longer term studies conducted concomitantly. The Panel emphasizes deficiencies in "continuing-baseline," long-term research, especially continental-scale studies in coastal waters and the atmosphere. In these media, there is more knowledge of environmental processes, but insufficient background information on historic or even present concentrations of many important chemicals.

We are also ignorant of the normal range of variations through which ecosystems and their components pass in coastal waters. Without knowledge of background concentrations or biological variation, we have a very difficult time assigning causes to small pollution impacts, because we cannot distinguish them from natural variations. Once we are certain that a measurement is outside the normal range for the system, we will also need to study the processes that have caused such effects and how pollutants interfere with those processes. Without knowledge of the normal range of variation, we are unable to determine unambiguously that something has happened to the system. Long-term observations, on the scale of decades, are necessary for the oceans and the atmosphere.

Anticipatory Studies

Long-term coastal ocean and rural atmosphere monitoring stations could serve as ''anticipatory long-term research'' vehicles for identifying actual environmental problems at an early stage. Past experience is an indication that surprises are likely, and it is our intent that monitoring programs exist side by side with research programs to explain the monitoring observations.

Data Quality

Because of the expense associated with collecting long-term environmental data, the Panel notes that compliance and monitoring data at existing hazardous waste sites and other industrial waste sites, such as Superfund sites, could be used for research if scientists were involved in the initial design of data collection programs. Data gathered for environmental management, whether for litigation, for determination of environmental trends, or for policy development, must be valid for its proposed uses. In addition to our concern about the precision and accuracy of measurements, we must assure that the data collected are appropriate to the needs the data must serve. Monitoring data could be used in answering research questions by careful initial design of sampling programs. Research questions concerning the transport of chemicals away from the site are, in turn, germane to regulatory concerns about risks posed by waste sites.

Assimilative Capacity Predictive Models

The Panel deliberated extensively over the usefulness and meaning of the concept of ''assimilative capacity'' of the environmental media to process waste loadings. We recognize that recycling and re-use, pretreatment, and process and product substitutions are the cornerstones of any waste management strategy. We note that the term ''assimilative capacity'' can have different meanings among disciplines. Many of these differences can be attributed to media characteristics. For instance, in containment media such as soils and some coastal estuaries, assimilative capacity refers to the amount of material added to a system before the system begins to leak, releasing the wastes to the outside environment. In dispersive environments, such as the atmosphere and open ocean, this physical sense of the term would refer to a rate of waste loading that is sufficient to allow waste concentrations to be diluted to the range of background variation outside of a mixing zone. Especially for soils, we know relatively little about assimilative capacity, with the exception of the assimilative behavior of soils with respect to certain metals.

Both of these transport-based definitions also may include, although they have not usually in the past, consideration of the extent to which these waste loadings are transformed by physical, chemical, or biological processes to other substances, the rates of transformation, and the degree to which these transformed substances accumulate in the environment.

Transport-based definitions of assimilative capacity require knowledge of geochemical, hydrologic, and transformation processes. Assimilative capacity has a further biological definition in engineering design of "the amount or rate of waste addition that does not cause unacceptable adverse impacts to the public health or to the environment." This effects-based definition of assimilative capacity requires additional information on long-term biological effects of pollutants on receiving media.

The determination of assimilative capacity in environmental management requires the development of predictive models. We believe that a useful first step for model formulation is to use simple models coupled with monitoring to permit model validation. Initial case studies should emphasize pollutants that exist primarily in one environmental state, whether gaseous, liquid, or particulate. In addition, compounds should be chosen that have similar residence times in each medium.

Institutional Arrangement Needs

Although we believe that an intermedia approach is necessary for proper waste management, we note that extant laws and regulations relating to environmental protection largely focus upon single media. Changing institutional arrangements to allow more of an intermedia perspective will present problems requiring long-term political and administrative research as well as research in the technical areas discussed above. Multimedia regulation may be possible by only slight modifications of existing institutions or may require extensive changes. In appraising alternative institutional means for achieving multimedia control, the feasibility of legislative and regulatory adoption and implementation must be determined. Better efforts are required to understand the political science and policy of environmental legislation. One method to improve our understanding of the efficacy and feasibility of multimedia control is to explore our historical experience with environmental protection at the legislative and administrative levels.

Institutional Research Needs

We believe that the sorts of intermedia, interdisciplinary, long-term research that we have discussed above are not adequately addressed by the current institutional structure. For instance, no academic institution has a curriculum that is adequate for the preparation of its students for tackling intermedia issues. Additionally, the measurement of substances in any medium over long time periods will be quite difficult, as evidenced by case histories of past attempts. Example are the long-term measurements of carbon dioxide in the atmosphere.

Improved institutional arrangements are needed to assure adequate funding, conduct research, and apply the research results to the solution of environmental problems. We have not identified what sort of institutional structure would be most appropriate. We have agreed that important attributes of the institutions conducting research are that they be interdisciplinary, nonpolitical, and capable of taking a broad, integrative view. Several panelists cite the EPA's Centers of Excellence or the National Center for Atmospheric Research as possible models.

RECOMMENDATIONS

In developing our recommendations, the Panel concurs with a seminal principle for research on media processes and their protection: "our environmental media—air, soil, and water resources—are vital to life on this planet and inextricably linked; our goal is to preserve our soil, air, and water resources and their productivity; these resources will always serve as an ultimate sink for wastes, though the amount and hazard associated with current waste inputs can be reduced; and protection of these resources is economically more reasonable than allowing contamination and then attempting restoration." We also believe strongly that the private sector should play an increased role in the improvement of existing monitoring techniques and protocols. We emphasize the importance of information transfer to environmental managers in both government and industry.

Integrated Approaches

We recommend the following research activities:

- intermedia transfer rates should be determined in an intermedia modeling/monitoring program that is focused on intermedia transport as a function of different physical parameters;
- results of intermedia modeling/monitoring programs should be used to develop screening procedures to determine for which environmental conditions and chemical characteristics it is necessary to consider intermedia transfer;
- effects of chemical interactions on transport processes should be studied; and
- there should be a linking of these transport processes to the biological effects seen in receptor organisms affected by exposure to the chemicals, whether they be humans, commercially important species, or indicators of the health of whole ecosystems.

Fundamental Processes

We recommend the following research activities:

- Research on transport and transformation processes should be conducted as part of site-specific field studies. Extrapolation to field situations from laboratory results has been extraordinarily difficult to date, and until more general principles are revealed, site specificity will remain a perplexing problem for soil/groundwater studies. We recommend that existing hazardous waste sites and other industrial waste sites and well-functioning Resource Conservation and Recovery Act (RCRA) sites be investigated as candidates for study. Given the expense of groundwater monitoring, we believe that research can most efficiently be coordinated with compliance monitoring at existing sites at an early stage of development of the sampling plan. Besides existing sites, older sites can also be explored similarly to the way archaeologists exhume ancient civilizations, reconstructing important processes occurring years earlier through the analysis of chemical residues. We recognize that there may be legal constraints on such a research approach because of such questions as liability, but we believe the idea is worth pursuing; and
- Atmospheric research should be aimed at understanding fundamental processes controlling tropospheric biogeochemical cycles on a global and continental scale. These processes include studies of source and sink processes for atmospheric gases, particles, and research on photochemical reactions.

Continuing Baseline Studies

We recommend the following research activities:

- For coastal waters, several small estuaries (for example, Buzzards Bay, MA, South San Francisco Bay, CA, Mobile Bay, AL, or Trinity Bay, TX) should be selected to determine the mass balance of contaminants while establishing the natural biological communities residing in these estuaries and the effects of contaminants on those communities. Mass balances should also be conducted on coastal areas that have historically received large waste loads. Improved collection of data about chemical release is necessary to do this. As yet, we have not been able to construct satisfactory contaminant mass balances for any estuary. This problem should be most tractable in small estuaries, where it is only necessary to consider relatively few processes; and
- For the atmosphere, long-term baseline measurements (at least 10 years) should be collected for the most important pollutants. These would include ozone, oxides of nitrogen and sulfur, and methane—on both a continental and global scale. Baseline rural atmospheric measurements, in particular,

are emphasized because they currently lie outside the range of responsibility of the present institutional structure. These data are needed to define the current state of the atmosphere, to shed light on important source mechanisms, and to allow determination of long-term trends.

Assimilative Capacity

We recommend the following research activities:

- Information should be developed regarding the amount of waste generated and disposed into each of the environmental media, the amount of reduction available under the above options, and whether exposure could be reduced by handling the waste in another medium. We also recommend that new methods be evaluated for their potential to detoxify municipal and industrial wastes. For example, wet-air oxidation, an aerobic thermophilic process, may be technically and economically feasible for treatment of concentrated organic wastes; and
- Regional evaluation methods and techniques should be developed for predictive purposes, including the development of single-media and intermedia models. There is an immediate need for methods to screen out inappropriate or outmoded waste sites and to verify pollutant transport models. Models appropriate for one set of uses or conditions may be totally inappropriate in a slightly different context. It is important to ensure that the model is valid for the uses that are required of it. We are concerned that the models not require unrealistically extensive data collection. The data collection requirements for many existing models quickly become impractical under normal field settings.

Institutional Arrangement Needs

We recommend the following research activities:

- Respective costs and benefits of multimedia control as a regulatory strategy should be characterized and appraised in comparison to single- or dual-media alternatives. Pollutants that can be more effectively controlled through multimedia control strategies should be identified. It may be that a multimedia approach is only necessary for a small number of cases;
- Alternative strategies of multimedia control should be characterized and appraised, with such strategies defined by (a) alternative goals and objectives; (b) alternative institutional means to achieve goals and objectives; and, (c) requirements for scientific and other technical inputs. In particular, there should be an evaluation of the usefulness of approaches such as that embraced in the Toxic Substances Control Act as a basis for regulating toxic and hazardous wastes by means of multimedia control. For instance,

case studies of EPA's Office of Groundwater and the evolution of the Groundwater Strategy, or of EPA's current acid rain activities, could yield valuable insights into administrative constraints to toxics integration efforts, because each of these potential subject areas involves toxic problems with multimedia dimensions.

LIST OF CONTRIBUTORS

Dr. Nathan Buras
Department of Hydrology & Water Resources
University of Arizona
College of Engineering
Tucson, AZ 87521

Dr. Yoram Cohen
Department of Chemical Engineering
University of California
5531 Boelter Hall
Los Angeles, CA 90024

Roger C. Dower
The Environmental Law Institute
1616 P Street NW
Washington, DC 20036

Dr. Raymond C. Loehr
College of Engineering
Cockrell Hall 8.614
University of Texas
Austin, TX 78712

Dr. Wayne A. Pettyjohn
Department of Geology
Oklahoma State University
Stillwater, OK 74078

Leslie Sue Ritts, Esq.
Morgan, Lewis & Bokius
2500 M Street, NW
Washington, DC 20037

Dr. John Teal
Biology Division
Woods Hole Oceanographic Institution
Woods Hole, MA 02543

Dr. Louis J. Thibodeaux
Hazardous Waste Research Center
Louisiana State University
3418 CEBA Building
Baton Rouge, LA 70803

Dr. Steven C. Wofsy
The Center for Earth & Planetary
Physics
Harvard University
Pierce Hall 100–A
29 Oxford Street
Cambridge, MA 02138

1
IMPROVING THE KNOWLEDGE BASE ON SURFACE WATER AND GROUNDWATER RESOURCES

Wayne A. Pettyjohn

INTRODUCTION

Historical Perspective

Major hydrologic investigations in the United States, other than for local water supplies, began with the exploration of the major waterways in order to provide insight into the development of western lands. Regional groundwater studies came into being largely following the Civil War. In the late 1800s to the mid 1930s both state and federal agencies conducted surveys to map and evaluate regional aquifers, such as the Dakota Sandstone. In the early twentieth century studies were also conducted on local aquifers, largely to determine well yields.

In 1935, Theis published his nonequilibrium well equation, which placed a new dimension on groundwater that has continued to the present time. The equation permits calculation of the transmissivity and storativity of an aquifer, both of which are required to estimate well yield and the extent of the cone of depression. From 1935 until the mid-1960s most hydrogeologists were involved with the examination and evaluation of individual wells and well field design. Most of the research dealt with improvements on the analytical techniques of well evaluation. For nearly a decade beginning around the mid-1960s, investigators developed computer programs to model the flow of groundwater. This marks another major advancement in the art and science of groundwater hydrology.

Until 10 or so years ago, most groundwater studies were involved, almost exclusively, with water quantity. Discussion of chemical or biological quality was usually limited to only a table or two of data. Due largely to the news media, pollution became a widely used term in the 1970s and this forced the scientific community to take a new look at water. Owing to their obvious nature, streams were cleaned up first in response to the development of new laws and technology, and this was followed, to some extent, by air pollution. When it was no longer legal to dump wastes into streams or burn them, the only apparent and cheap disposal method was to put them on or in the ground. Little thought was given to the possibility of groundwater contamination.

Particularly following World War II, the chemical industry began to develop scores of new organic compounds that were used for thousands of purposes, such as solvents, detergents, plastics, synthetic fibers, and pharmaceuticals to mention only a few of an exceedingly large number. The residues were thrown away. Because of the improved economy nearly every family owned a car and gasoline stations appeared on almost every corner. Moreover, agricultural chemicals began to be widely used to provide the crops necessary to feed a hungry world.

Only within the last few years have we begun to recognize the problems that will be faced in the coming decades. Many of the products that we used, stored, or threw away migrated through the "living filter" of the unsaturated zone and found their way to the groundwater reservoir, contaminating it with a wide variety of compounds about which we know little. Millions of underground storage tanks, which have a design life of about 20 years, are now leaking, many agricultural chemicals that were thought to be immobile in the subsurface are appearing in groundwater over wide areas, and the leachate from landfills and dumps has contaminated domestic, industrial, and municipal wells alike. Recognizing these problems, regulatory agencies and the courts have developed design guidelines, laws, and rules to protect the environment. New initiatives are being developed. Resultingly, the scientific community is now required, commonly by court order, to provide solutions to problems that were never even considered previously. We have no good data base to provide sound solutions to many of our emerging problems and, in many cases, we are using nineteenth-century techniques to solve twenty-first-century problems.

Need For Long-Term Research Initiatives

Most of the water quality research has been funded by the U.S. Environmental Protection Agency (EPA). The agency has also exerted a strong control on the topics covered, which is reasonable because they require the results for their regulatory role. This concept has not been without problems, however. A number of the research initiatives were and are of an ephemeral nature, perhaps brought about by high visibility in the news media coupled with public, legislative, and legal pressure. This usually results in short-term interest.

Another significant impact on research is brought about by policy makers, a great many of whom have little or no technical expertise. Resultingly they have little understanding of the complexity and interactions involved in many environmental matters. One example is the Ground-Water Protection Strategy, which is slowly making its way through EPA. As most technical people are well aware, there is no data base or body of technical information that would allow the agency to accomplish many of the objectives that are proposed in the Strategy. Although the ideas behind the Strategy are sound, the lack of recognition of the technical problems in achieving the desired results indicates only a superficial understanding of the magnitude of the problems faced by the agency and state personnel. Moreover, the Strategy does not appear to adequately recognize the need for research. Physical, chemical, and biological values and parameters, on which predictions and engineering designs must be based, are largely little understood, misunderstood, ignored, or merely assumed. Only research can fill in the blanks, but the research must have sufficient longevity to achieve results, funding must be consistent and at an adequate level, and contract renewals must be handled in a timely manner.

Research And Technology Transfer

Research and technology transfer must walk hand in hand and the latter is equally as important as the former. Far too many research reports serve only as dust collectors. The reason, of course, is that some are far too long and complex to be understood and appreciated by potential users. Moreover, only a few individuals are aware of research activities, even within the same agency. In fact, many research proposals and grants are almost identical but they are funded by different units in the same agency, none of whom are aware of what the others are doing. This is clearly a communications gap that can be exceedingly costly.

A mechanism must be made available to provide users with new ideas, techniques, and procedures. Esoteric research reports need to be rewritten in simpler, condensed versions. Seminars, workshops, and manuals offer one potential solution.

SURFACE WATER AND GROUNDWATER RELATIONS

Sources Of Streamflow

All the water in a stream flowing past a certain cross-section is called runoff. Runoff consists of precipitation that has fallen directly into the stream channel, water that has flowed over the ground to the channel (surface runoff), effluent, and groundwater that has flowed into the stream (groundwater runoff). Many streams owe much of their flow to groundwater runoff; the reason a stream flows during dry periods is because of discharge from the groundwater reservoir. In addition, the chemical quality of a stream during dry periods closely reflects the composition of groundwater in the zone of intensive circulation. In many respects a stream is nothing more than an exceedingly long, very shallow, horizontal well. Many hydrogeologists consider runoff to be no more than rejected groundwater recharge.

Water Resource Evaluation

Although not widely recognized, the interrelations between surface water and groundwater are of great importance in both regional and local hydrologic investigations and a wide variety of information can be obtained by analyzing streamflow data. Most commonly the surface-water investigator deals with stream hydrographs, channel characteristics, geomorphology, or flood routing. Although the hydrogeologist may evaluate induced infiltration into a streamside aquifer, he is generally more interested in aquifer characteristics, such as permeability, thickness, boundaries, and well yields. Many hydrologists tend to ignore the fact that, at least in humid and semiarid regions, groundwater runoff accounts for a significant part of a stream's total flow.

Evaluation of the groundwater component of runoff can provide important and useful information regarding regional recharge rates, aquifer characteristics, and groundwater quality, and indicate areas of high potential yield to wells. Moreover, interruption of the relation between groundwater and surface water can produce substantial and commonly unsuspected consequences, such as the rise of water levels in wells as a newly constructed reservoir begins to fill. Clearly a much broader picture needs to be viewed in any large-scale water development scheme.

LONG-TERM RESEARCH NEEDS—GROUNDWATER QUANTITY

Introduction

Until the past few years, most groundwater studies were involved almost exclusively with well yield and well field management. Despite the long years of study and considerable progress, much still remains to be learned. Nonetheless, there has been far more research, information transfer, and technique development than is the case with groundwater quality. In the case of groundwater hydraulics, however, it is suspected that the scientific community will need to develop a new heading—a heading that, to some extent, returns to the basics.

It appears that many of the fundamentals of groundwater have been forgotten, ignored, or were never known.

Groundwater Recharge

Groundwater recharge is the process whereby water on the surface infiltrates to the groundwater reservoir, replacing the quantity that is discharged to streams, springs, seeps, and wells. There is a long-term balance between groundwater recharge and discharge and, consequently, the water level maintains an average level, at least in unstressed areas. Where the pumping of groundwater exceeds the long-term recharge rate, water levels decline. Substantial declines increase the cost of lifting the water, reduce well yields, and in certain circumstances result in land subsidence, as is evident in numerous places such as Mexico City, Venice, Tokyo, and several areas in California, Arizona, and Texas, among others.

Natural Groundwater Recharge Research

The sustained yield of a well field depends on the quantity of natural recharge that is captured within the cone of depression. Natural recharge also exerts a strong influence on runoff. Furthermore, it generates and also dilutes leachate plumes and forces them to migrate deeper into the subsurface. EPA is presently developing a system for aquifer classification and one of the input parameters is recharge.

Only a modest amount of research dealing with natural recharge has been conducted. Generally hydrologists assume that the rates range from about 12 to 20 percent of precipitation, at least in humid and semiarid regions. Some studies, however, clearly show that natural recharge varies throughout the year and from one year to the next. A major control appears to be antecedent precipitation. Because natural recharge has such a profound effect on many parts of the hydrologic cycle, it should be studied in far greater detail. The studies will require several years for completion.

Artificial Groundwater Recharge Research

In many parts of the world groundwater pumpage far exceeds the rate of natural recharge. Consequently, water levels are locally declining at alarming rates. This is particularly true in some arid regions. In the vicinity of Torreon, Mexico, which lies about 600 miles north of Mexico City, water levels are declining at a rate of 1 to 3 meters/year due to the withdrawal of water for irrigation. Similar situations occur throughout much of Mexico's desert region. In the not too distant future the situation may well become critical and of international concern because these irrigated areas are becoming the breadbasket of our southern neighbor.

One can only speculate on the turmoil brought about by a food shortage in a country already beset with nearly unmanageable economic problems.

Similar decline problems, but fortunately of lesser magnitude, face many of the irrigated regions in the United States. The High Plains aquifer, stretching from Nebraska to Texas, is being dewatered to such an extent in some areas that the water level has declined more than 100 feet. On the other hand, most severe water-level decline problems are of a local nature, being restricted to municipal and industrial well fields.

Artificial groundwater recharge is a potential solution to aquifers that are being dewatered. Artificial recharge is a means of augmenting the natural infiltration of surface water into a groundwater reservoir at a rate that vastly exceeds that which would naturally occur. It includes a variety of methods, such as wells or other specialized construction, water spreading, or changing natural conditions. Various artificial recharge techniques have been used throughout the world for more than 200 years.

Despite the long history of artificial recharge, much still remains to be learned. Studies of this nature must be continued over a period of many years, in a variety of climatic and geologic settings, and supported by a substantial budget.

Evaluation Of Units Of Low Permeability

Until recently rocks of low permeability were largely ignored in groundwater investigations because they were assumed to form confining units through which water could not flow. Eventually it became apparent that aquifers enclosed within these units provided far more water than the aquifer could physically contain. It was then realized that confining beds store vast quantities of water that slowly leak from them to replenish the more permeable strata where water is held under a lower pressure.

Investigators evaluating underground coal gasification in western states during the last seven or eight years found that the usual groundwater evaluation techniques did not provide adequate answers. The reason is that the normally used equations were designed for flow through porous media and not slightly fractured rocks of very low permeability.

It has become common practice to site waste disposal facilities on strata of low permeability in order to protect groundwater supplies from contamination. Many individuals seem surprised when leachate is discovered a considerable distance from the site. Again the reason is that the water-transmitting characteristics of the confining beds are poorly understood and techniques to evaluate aquifers are used, inappropriately, to assess them.

Both pure and applied research efforts are needed to develop methods to evaluate and characterize units of low permeability. Investigations of this nature are particularly timely because of the need to find adequate hazardous waste disposal sites.

Significance of Fractures And Macropores

Although long recognized, only recently has the significance of fractures and macropores in aquifers and confining beds been realized. These highly permeable conduits permit the rapid transfer of water through rock units. Furthermore, flow through fractures rapidly transmit both chemical and biological materials that lead to groundwater contamination and possibly to outbreaks of waterborne disease. It also appears that rocks of low primary permeability, such as shale and glacial till, contain an abundance of fractures. This is particularily important because these are the types of strata in which landfills and hazardous waste dumps are generally located. Consequently, instead of retaining leachate the fractures permit rapid transfer of fluids along paths that are not readily traceable.

Macropores are large openings in the soil caused by burrowing organisms, rootlets, desiccation cracks, and construction activities. Many are ephemeral in nature, being closely related to soil composition and moisture content. Because of flow through macropores, groundwater quality can change abruptly following periods of precipitation. Flow occurs even if there is a soil moisture deficiency. It appears that waterborne disease outbreaks may be more prevalent during the spring and fall, periods when macropores are most common. The large openings offer little if any filtering action and are likely to quickly transfer pathogens to a source of water supply.

More research needs to be conducted on the significance of fractures, their detection, and temporal nature.

Evaluation of Runoff to Estimate Groundwater Conditions

As discussed previously, a great deal of groundwater information can be obtained by analysis of streamflow data. Information on potential well yields, regional effective recharge rates, aquifer coefficients, and chemical quality are usually obtained by means of test drilling, which is both expensive and time consuming. On the other hand, streamflow information is readily available, covers a wide geographic area, and represents many years of collection.

There has been no long-term research effort on the development of techniques to more effectively utilize streamflow data for groundwater purposes. However, since streamflow originates from large areas, it can be used to obtain regional evaluations quickly and with very modest cost.

Development Of Regional Evaluation Techniques

To some extent streamflow data can be used to obtain regional hydrologic evaluations, but satellite imagery offers a broad overview and computer flow models aid in understanding the hydrologic system. New methods are needed to combine the various existing techniques in order to obtain maximum information.

Measurement Improvement Techniques

It seems incredible in this period of space-age technology that many if not most of the hydrologic measuring techniques were developed a half century or more ago and have limited accuracy. Examples include the measurement of the areal distribution of precipitation calculation of streamflow, and the depth of water in wells. The latter is a particularly vexing problem due to the considerable depth to water in many wells and well construction that prohibits the introduction of a tape into the well bore.

New tools are needed to measure both large and small areas. Satellite data might be more effectively used to calculate precipitation, particularly if the imagery can be calibrated. Microelectronics offers a possible solution to runoff measurements. Water levels in wells might be accomplished with geophysical techniques that transmit a signal through the well or well casing with the return signal being recorded at the surface.

LONG-TERM RESEARCH NEEDS—GROUNDWATER QUALITY

Introduction

Because of the new awareness of groundwater contamination, particularly contamination by organic compounds, investigators are now being called on to answer questions that have never before been asked. Furthermore, the interrelations are vastly more complex than has been anticipated. For example, to understand the migration of organic compounds in surface or groundwater and to make predictions requires communication between microbiologists, soil scientists, hydrologists, organic chemists, engineers, and mathematicians, among many others. Yet communication is difficult at best because of wide differences in technical vocabulary, analytical procedures, and techniques.

Researchers at EPA's Robert S. Kerr Environmental Research Laboratory in Ada, OK prepared a Ground-Water Research Strategy (EPA, unpubl.) that was divided into three major topics, each of which contained a number of elements. The major topics are: (1) hydrologic processes, (2) abiotic processes, and (3) biotic processes. The Research Strategy was developed over a period of many months and represents the input of some of the most active researchers in the world. This document is the basis for the following discussion.

Research Needs in Hydrologic Processes

Hydrologic processes include the influences of physical and chemical characteristics of the subsurface environment on the movement of natural and contaminated fluids through it. These processes also include peripheral factors, such as the interaction of groundwaters and surface waters and the influences of man's activities on groundwater recharge, discharge, movement, and quality. It is necessary to thoroughly study these processes on both macroscopic or regional levels as well as microscopic or site specific levels.

Subsurface Methodology

To gain an understanding of the hydrologic processes and physiochemical features that control water movement, many sampling and analytical methods must be developed or refined. These methods include (1) collection of subsurface rock and fluid samples that are neither disturbed nor contaminated, (2) sample preservation and analytical techniques that permit highly accurate measurement of the physical and chemical properties of the sample, and (3) techniques that rapidly and economically define variations in physical and chemical properties of the subsurface.

While most of these topics have been examined for many years, the difficulty in dealing with complex subsurface systems and the need to focus on minor variations of the movement of contaminants that occur in the parts per billion range have made even more vigorous research in this area essential.

Aquifer Characterization

A broad term, aquifer characterization denotes the complete physical, chemical, and biological definition of an aquifer. It includes the techniques listed under Subsurface Methodology as well as the use of statistical principles for efficient sampling methods and data interpretation. Basic research regarding the fundamental geologic processes and occurrences that led to the creation of each aquifer is also a key element. Additionally, demographic influences on aquifer behavior are required topics of study in aquifer characterization. Research of this type is site specific.

Contaminant Plume Delineation

The geometry of contaminant plumes in groundwater are determined by drilling, groundwater sampling, geophysical techniques, and physical or computer simulation models. Of particular importance in studying contaminant plumes are the dynamics of mixing/dilution and attenuation/retardation. Mixing/dilution is closely related to dispersion in an aquifer, which is a characteristic that for all practical purposes can not be adequately measured. Nonetheless, most predictive models require dispersion as an input. Attenuation/retardation processes are currently being studied for a limited number of chemical and biological contaminants under a few field and laboratory conditions. Field efforts must continue to focus on rapid, accurate, and economical means of plume identification, while laboratory efforts must continue to attempt to elucidate the mechanisms of plume attenuation. Furthermore, major advances need to be made in order to properly scale the magnitude of laboratory findings with field situations.

Spatial Variations in Hydraulic Properties

Most computer models and traditional analytical methods assume each aquifer system is homogenous or near so. Since this is the exception rather than the rule, methods need to be developed and/or refined to locate and accurately evaluate abrupt changes in hydraulic properties. Conventionally, methodologies have been employed to interrelate the distribution of hydraulic properties to readily measureable field parameters, as for example static water levels. Some of these are inherently geological, while the remainder are purely theoretical. The advances required by today's problems may be in the form of statistical approaches, geophysical techniques, or remote sensing, and will come only with full integration of both geologic realities and theoretical considerations governing hydraulic properties.

Unsaturated Flow

The influence of unsaturated flow conditions on the attenuation of contaminants is as yet only poorly understood. While the fundamentally retarding impact of unsaturated conditions on fluid movement is somewhat understood, methods to predict that impact on a detailed level are notoriously underdeveloped. A great deal of the difficulties in developing satisfactory methods arises from the tremendous variability and complexity of the physical and chemical properties of unsaturated flow zones in the subsurface.

Site-Specific Homogeneous Aquifers

Some aquifers are either relatively homogeneous or are studied on such a scale that localized variations in hydrogeologic properties can be ignored. The best, most economical field methods that can be used in these situations to obtain a good approximation of aquifer properties include conventional drilling, hydraulic testing, and water sampling and analysis. Improvements in aquifer testing, data analysis, in situ measurements, and numerical simulations, however, are still needed because so few studies have been conducted at the level required for contaminant migration investigations.

Site-Specific Inhomogeneous Aquifers

Field data from investigation of aquifers with abrupt changes in thickness and hydraulic properties are the most difficult to evaluate and model with any degree of accuracy. Traditionally, information is based on drill hole data and these may be far too few to develop even a crude understanding of complex hydrologic sites. Field methods that can be used in conjunction with test hole data must be developed. Statistical analyses and innovative mapping techniques, which effectively reduce drilling and data acquisition requirements, are desperately needed.

Mathematical Description of Water Movement

Mathematical models offer the most powerful and convenient approach to prediction and management of fluid movement in the subsurface. Both analytical and numerical mathematical models have become highly advanced in recent years, but field applications of these models have routinely lagged far behind model development. Field applications are greatly needed to calibrate and verify existing models. Because of the site-specific nature of the physical and chemical factors affecting fluid movement, it is still nearly impossible to create generally applicable models and, therefore, much remains to be done in this area. Screening methods and benchmark problems that can be used to assess the relative merits of various models must be developed.

Groundwater Research—Abiotic Processes

A number of physicochemical processes attenuate the movement of pollutants in the subsurface. These processes, such as sorption and chemical reactions, may slow or possibly enhance movement, stimulate microbial degradation, and facilitate transformation of the pollutants. An understanding of the mechanisms and extents of these processes will enable mathematical modelers to better predict the transport and fate of such chemicals in the subsurface, the potential impact of these chemicals on groundwater resources, and the probable benefits of rehabilitation efforts. The research elements enumerated under this heading focus on key topics to advancing the knowledge of subsurface abiotic processes.

The collection of subsurface material and groundwater samples for monitoring or experimental purposes is crucial to developing an understanding of the behavior of pollutants in the subsurface. Additionally, methods to evaluate treatment and rate of interactions between the pollutants and the solid phase need refinement. Recent advances in pump design and well drilling technology have enhanced the prospects of collecting minimally disturbed samples, but more development is needed, particularly with respect to volatile organic compounds and pollutants sensitive to redox conditions, in approximating field conditions in laboratory interaction/reaction experiments, and in monitoring pollutant behavior in the field.

Identification/Characterization of Soil Constituents with Respect to Pollutant Behavior

Pollutant movement through the subsurface will be mitigated by the nature of the solid constituents and the coatings on the surfaces of these constituents. Well-developed relationships for sorption of hydrophobic organic compounds to soil organic matter and for certain inorganic ions to soil mineral constituents already exist. The degree of interaction of organic and some inorganic pollutants with such soil components as clay minerals and hydrous metal oxides still remains to be defined. Research needs include: solid phase constituents, knowledge of sample storage effects and test variability, and screening techniques to determine the degree of pollutant-solid phase interactions.

Interactions and Relationships between Subsurface Constituents

Very little is known about the extent to which clay and soil organic matter and sesquioxides combine and how the combined material affects pollutant behavior. The experiments that have been done to date would indicate, for example, that a chemical combination of some soil organic matter with selected clays would have a much greater adsorption capacity for some pollutants than either component alone. Chemically combined compounds of clays and soil organic matter constituents are very difficult to distinguish from mixtures of these. The techniques available at the present time revolve around the use of transform infrared spectroscopy, wherein the bonds between organic matter and the mineral fraction can be determined, possibly enabling distinguishing the combined material from individual mixtures.

Spatial Variability of Soil Properties

Many soil properties, including organic carbon, clay content and ionic strength, and composition, have been shown to influence pollutant fate in soils. Knowledge of these properties and their temporal and spatial variability is essential in accurate prediction of solute transport and in development of field scale sampling strategies. The limited efforts to date in defining spatial variability have been largely confined to generating information on hydraulic and physical properties in surface soils. Research efforts require expansion of previous efforts plus inclusion of vertical profiles of chemical soil properties.

Spatial Variations in Natural Groundwater Quality

Groundwater changes its composition in its path from recharge to discharge. The details of such changes, however, both horizontally and vertically, have seldom been documented. In addition, there may be seasonal changes impressed on these longer-term changes. Cyclic variations, often tied to critical changes of pH or Eh, have been observed in different areas of groundwater systems. Collection of in situ data on important system parameters, such as subsurface redox conditions, information critical to understanding chemical behavior of subsurface pollutants, is virtually impossible with current techniques. However, it is conceivable that common geophysical borehole techniques, such as those applied in the petroleum industry, could be applied to the measurement of subsurface environmental parameters. Information on the natural variations in groundwater quality coupled with insight into the role that such quality has on pollutant behavior will ultimately facilitate a greater understanding of pollutant mobility in the subsurface.

Relationships between Soil Constituents and Natural Groundwater Quality

A fundamental premise in groundwater geochemistry is that the composition of the groundwater is largely a result of interactions with solid materials in the subsurface over which the water passes in its path from recharge to discharge. These interactions will also greatly influence the redox potential of the ground water. While simple inorganic chemical equilibrium reactions serve to explain and predict behavior of the high concentration salts in groundwater, methods for predicting trace level chemical changes are still awaiting development. Depending on what subsurface constituents are actively involved in solute-surface interactions, system parameters such as pH and Eh (redox potential) will likely have a direct impact on the surface chemistry involved in such interactions. The role of system parameters in influencing soil formation and stabilization, weathering of geologic materials and aqueous trace chemical concentrations is fundamental research requiring a much clearer definition.

Interaction of Soil Constituents and Pollutants

The interaction between pollutants, including viruses, and soil constituents is dependent on three primary factors: the nature of the pollutant; the composition of the soil; and the physical-chemical conditions of the system. The relative significance of soil components in interacting with inorganic and organic pollutants and viruses needs considerable refinement. Knowledge on the significance of system parameters, Eh, pH, soil moisture and organic carbon content, as well as others, needs to be extended to the whole range of pollutants and soil components. Available information on pollutant characteristics should be focused to ascertain the relative significance of these characteristics in influencing the pollutant's behavior in the subsurface. Finally, the nature and significance of these interactions must be determined as they occur in the field.

Nature and Extent of Abiotic Reactions in the Subsurface

Subsurface environmental factors favor the occurrence of abiotic reactions, such as hydrolysis and redox reactions, in addition to, or possibly in conjunction with, microbial activity. Three factors in particular: high surface areas of clays and metal oxides; reactive moieties in soil and dissolved organic matter; and the slow flow of groundwater providing long contact times would function to enhance abiotic reactions. Laboratory methodologies for reproducing the conditions necessary to investigate the nature and extent of such reactions and the environmental conditions that influence them remain to be devised, and field capabilities in this area are currently quite limited. Additionally, the interrelationships between surface, abiotic, and biotic reactions in the subsurface will need to be defined before the degradative processes can be understood.

Effects of Large Amounts of Pollutants

Large amounts of pollutants can be introduced into the subsurface as a result of accidental spills, breaches in waste disposal pond or landfill integrity or unlawful disposal practices. Movement of large amounts of pollutants through the subsurface will likely produce disruptions in natural biotic and abiotic processes and changes in physical-chemical subsurface characteristics (the potential for clay alterations by organic solvents has been demonstrated). While numerous examples of the introduction of large amounts of pollutants into the subsurface are cited in the literature, very little research has focused on the effects from such occurrences or the development of information to minimize the consequences.

Extent of Dispersion and Diffusion

The dispersion term of the solute transport equation, describing the distribution or shape of a noninteracting chemical front or pulse moving through porous media, represents a combination of molecular diffusion and hydrodynamic dispersion. In practice, however, this term is often used as a "catchall" for other poorly understood or defined processes. Field measurement of dispersion is an extremely difficult task. Future research needs to focus on the relationships of dispersion to the characteristics of the fluid, the porous media, and the environmental system and on the relative significance of dispersion and other transport properties, such as sorption, mass flow, and degradation, especially for trace level pollutants.

Extent and Rate of Volatilization from the Subsurface

It is likely that volatilization from the saturated and unsaturated zones accounts for a major fraction of trace organic losses from groundwater systems. Volatilization kinetics of organic compounds from water in a beaker are reasonably well-defined, but the extension of this information to subsurface systems has not been well-documented. Field studies with relatively nonvolatile pesticides have shown that soil characteristics influence volatility. Clearly, the theory of volatilization from partially water saturated soils needs development. Field methods development followed by field studies on trace organics will provide information essential to optimizing both the theory and the predictive basis of volatilization.

Extent and Rate of Soil Attenuation of Pollutants

A predictive understanding of the abiotic processes that influence pollutant movement in the subsurface necessitates knowledge of the kinetics and the magnitude of these processes. Current knowledge in this area is limited to certain inorganic ions on surface soils and to hydrophobic organic pollutants in relatively high organic carbon soils, those systems where the interactions are of the greatest magnitude and are most easily defined. Future research will expand to cover the range of pollutant types and the variety of subsurface systems encountered in the environment. The ultimate measure of this research will be its ability to provide, directly or through extrapolation, the information needed to evaluate any pollutant transport problem encountered in the subsurface.

Mathematical Description of Mass Transport Properties

Mathematical description of mass transport properties of subsurface systems are essential to the proper design and execution of research efforts and, ultimately, to the accurate prediction of pollutant movement in the subsurface. Such models currently represent an integration of several poorly defined topical areas, including numerous hydrologic and chemical relationships that still require extensive research to develop a functional level of understanding the modeling process. Mathematical description will be a topic of continual refinement, with the accuracy of model predictions improving as subsurface relationships become better defined.

Groundwater Research Strategy Elements Biotic Processes

The extent and nature of microbial activity in the subsurface must be understood to effectively develop groundwater quality management practices. Just as microorganisms are by far the most important agents that transform organic compounds in surface soils and waters, an active subsurface biota could have a profound effect on the movement and persistence of pollutants in subsurface materials and waters. On the other hand, the absence of appreciable microbial activity against pollutants in subsurface regions will signify that the fate of pollutants in groundwater will be controlled by abiotic processes only. The research elements discussed in the Biotic Processes category delineate key research topics for developing necessary information on subsurface biotic processes in order to predict the transport, fate, and impact of pollutants in the subsurface and to develop control and remedial technology for groundwater quality.

Subsurface Methodology

The major deterrent in the study of subsurface biological processes is a lack of reliable and/or inexpensive methods for examining the subsurface. Considerable progress has been made in recent years in developing methods for obtaining uncontaminated samples, in developing new techniques and procedures for characterizing subsurface biota, and in developing technology for determining how geological processes affect pollutant transport and fate. However, improvement is needed in all of these areas. Currently coring procedures are slow, cumbersome, expensive, and only work in certain types of subsurface materials. Whether cores should be handled under anaerobic conditions is not definitely known. The proper procedures for sampling well microflora have not been established, nor has it been determined what the value of such samples is. Methods for measuring in situ activity and conditions affecting that activity have been primarily limited to the use of physical models which are in the first stages of development and evaluation. Techniques and procedures for determining biomass, measuring important environmental conditions such as redox potential, examining ecological interactions, and determining the specific metabolic activities and capabilities of subsurface microflora are in the developmental stages and need continued improvement.

Nature and Vertical Extent of Subsurface Biota

The presence or absence of a subsurface biota could have a profound effect on the persistence of pollutants in groundwater. In regions of the subsurface that lack microbes in sufficient densities, the transport and fate of pollutants will be controlled by abiotic processes and will, therefore, be easier to predict. Current information indicates that the deeper subsurface environment is not sterile and that there may be considerable spatial variation in microbial populations both from a qualitative and quantitative standpoint. These conclusions, however, are based on a very limited number of studies and far more information on the distribution, density, and nature of organisms in the subsurface, both above and below the water table, is needed. Further studies should emphasize correlating the occurrence and activities of organisms with the geological and mineralogical properties and the environmental conditions of the environments where the organisms exist.

Effect of Interactions Between Subsurface Organisms

A divergent microflora is known to occur in at least some regions of the subsurface and may have a significant impact on the fate of many groundwater pollutants. Almost nothing is known concerning the interactions of subsurface organisms with each other or with surface organisms which enter that environment. In surface habitats such interactions can be extremely important in determining the fate of chemical or biological pollutants. For example, a microbial species with the ability to degrade a particular compound may not be able to compete for required nutrients with other organisms. Some microbes are dependent upon the by-products of other microbes' metabolisms. Whether such interactions are important in determining the fate of groundwater pollutants is unknown.

Nature and Extent of Biotic Reactions in the Subsurface

At present little is known about biodegradation of organic pollutants in the deeper subsurface. Current knowledge indicates that the microflora of shallow regions of the saturated zone can degrade certain organic pollutants, but cannot degrade other organic pollutants that are easily degraded in surface soil. Additionally, microbes from different sites vary in their ability to degrade particular organic pollutants.

A limited amount of pollutant biodegradation research has been done using the indigenous flora of the subsurface and foreign pollutants; virtually no work has been accomplished in situ. These results have indicated that the potential for significant biodegradation of a number of compounds exists. However, the question remains unanswered as to the degree these reactions will proceed in the subsurface. The limitations have not been clearly defined to be either thermodynamic (energy-limited) or kinetic (rated-limited) nor has the extent of adaptation and cometabolism been investigated in detail. Very little is known concerning degradation by-products or whether degradation processes can be manipulated.

Effect of Pollutants on Microflora

Little effort has been expended to determine what effect pollutants have on indigenous subsurface microflora. Such effects may be environmentally undesirable. For example, a recent study on the fate of organic pollutants in high rate land treatment systems shows that low concentrations of introduced organic compounds temporarily stopped nitrification. It is very possible that certain pollutants will be toxic to subsurface microbes and stop them from degrading pollutants which are not toxic. Another consideration is the mutagenicity of introduced pollutants, both to indigenous microflora and to introduced organisms, especially pathogens.

Effect of Geochemical Parameters on Biological Activity

The subsurface environment differs considerably from normally studied microbiological habitats found in surface waters or sediments. Little is known concerning environmental conditions in subsurface habitats and the effect such conditions have on biological activity and the biotic transformation of pollutants. Important factors governing the extent and/or nature of biological activity include: (1) the concentration and utility of electron acceptors; (2) the concentration and availability of essential nutrients; (3) the oxidation-reduction potential; (4) the pH; (5) the ionic composition; (6) the availability of water; (7) the temperature; (8) the hydrostatic pressure; (9) the nature of the solid phase; and, (10) the nature of the pore space. All of the above factors interact with each other to influence the activity of organisms in the subsurface.

Rates of Biological Transformations

Biodegradability and the associated kinetic relationships must be determined prior to development and use of mathematical models for predicting the movement and fate of pollutants in the subsurface environment. Volatilization and sorption must be determined to predict biodegradation since a mass balance is required. Reaction kinetics must be determined for modeling. Studies are required to determine: (1) the effect of concentration of pollutant on the rate law for transformation; (2) the effect of concentration of pollutant on the density of microbes active against that pollutant; and, (3) the correlation between the rate of transformation of the pollutant and the density of viable microbes, or the concentration of some biochemical constituent of the microbes used as an indicator of biomass or nutritional state. Current information is scant. Mathematical modelers tend to assume that biotransformations are first order. Detailed studies of biotransformation kinetics in material from the deeper subsurface are almost nonexistent, however, work on NTA, chlorobenzene, and dibromochloromethane suggest zero-order kinetics, at least at concentrations of 1–30 mg/1.

Mathematical Description of Biotic Kinetic Processes

To use laboratory and field information about subsurface biotic reactions to predict the fate of pollutants requires the development of mathematical submodels that describe the kinetics of biological transformations in the subsurface, and which can be incorporated into more sophisticated mass transport models describing water movement and abiotic attenuation of pollutants. Successful development of a biotic processes submodel will be dependent upon the determination of biotransformation kinetic information. Current mass transport models are deficient in handling biotic processes.

SUMMARY AND CONCLUSIONS

Groundwater is a major source of supply for domestic, municipal, industrial, and agricultural uses. The importance of this resource has long been ignored and, resultingly, hydrologists are now being asked questions that are unanswerable because the required body of technical information is not available. The data base can only be developed by a long-term, energetic research effort that is adequately funded. Priorities must also be established and, in many cases, these must be based on political and economic realities.

In the technologically advanced nations, the population seems to be most concerned with long-term health effects that might be brought about by drinking water contaminated with synthetic organic compounds. Except for local areas of excess water use and water rationing, little thought is given to adequacy of supply or water treatment. At the other end of the spectrum are many of the lesser developed countries where the only concern is finding enough water to survive from one day to the next. In these areas organic compound and heavy metal contamination of water supplies is of no concern whatsoever because the population suffers from an inadequate water supply and is decimated by waterborne diseases. Somewhere in between are those regions and countries, such as Mexico, where an advancing technology and, until recently, an improving economy has resulted in twentieth-century problems.

Many of the developing countries are attempting to become self-sufficient, particularly in regard to their food supply. Commonly this is based on irrigated agriculture, the regions are generally arid or semiarid, and consumptive water use is excessive. Resultingly, water levels decline rapidly and the aquifer may be near total depletion at the same time the region becomes dependent on it.

Water research activities in the United States probably first should be devoted to the development of an information base that would permit regulatory agencies to formulate adequate guidelines to protect this valuable resource and the general public from adverse health effects brought about by the long-term consumption of low levels of organic contaminants. This aspect would include a considerable amount of pure research. Second on the priority list would be investigations that deal with aquifer properties, such as development of techniques to evaluate rocks of low permeability, and water management. The final priority would include the formulation of methods to adequately describe the regional nature of ground water and surface water.

It is essential that all research activities be closely tied to a system of information transfer. A major fault of most researchers is the inability or unwillingness to prepare reports that can be applied to actual situations by field personnel.

BIBLIOGRAPHY

Britton, Gabriel and Charles P. Gerba (ed.), Groundwater Pollution Microbiology. New York, NY: John Wiley and Sons, 1984.

Deutsch, Morris, D. R. Wiesnet, and Albert Rango (ed.), Satellite Hydrology. Minneapolis, MN: American Water Resources Association, 1981.

Freeze, R. A. and J. A. Cherry, Groundwater. Englewood Cliffs, NJ: Prentice Hall, 1979.

Johnson, R. H., Relation of Ground Water to Surface Water in Four Small Basins of the Delaware Coastal Plain. Delaware Geological Survey Report of Investigation 24, 1976.

Pettyjohn, Wayne A. and Roger Henning, Preliminary Estimate of Ground-Water Recharge Rates, Related Streamflow, and Water Quality in Ohio. Columbus, OH: Ohio State University Water Resources Center, 1979.

Pye, Veronica, Ruth Partick, and John Quarles, Groundwater Contamination in the United States. Philadelphia, PA: University of Pennsylvania Press, 1983.

Siegal, Barry S. and Alan R. Gillespie (ed.), Remote Sensing in Geology. New York, NY: John Wiley and Sons, 1980.

U.S. Environmental Protection Agency, Proposed Ground Water Protection Strategy. Office of Drinking Water, Washington, DC: EPA, 1984.

U.S. Environmental Protection Agency, A Ground Water Research Strategy. Ada, OK: EPA, unpubl.

2
PROTECTION OF SURFACE WATER AND GROUNDWATER RESOURCES

Nathan Buras

INTRODUCTION

One can hardly conceive life on this planet without water. In the last hundred years, man has tried to store water in times of abundance so that it will be available in times of scarcity, and to transport some of the water thus stored to points of demand. Since the end of World War II, issues related to water resources development and utilization have been recognized as a scientific discipline grounded in basic hydrologic studies and supported by rigorous mathematical analysis. Much intellectual effort was spent in understanding natural hydrologic processes in order to devise solutions to problems raised by the ever increasing demands for water. During the decade of the fifties, the main concern of water resources scientists, engineers, and managers was to provide adequate quantities of water primarily for irrigation and also for domestic, industrial, and other uses. Water quality was of little concern.

The development of river basins in the United States during the last three decades in response to growing demands for water by all sectors of the economy—which strived for increased productivity—highlighted an additional set of problems that were latent before. Water, it turned out, was not only a part of the product (such as irrigated agriculture), but it was also a medium for transporting wastes away from the sites where production processes took place. The wastes carried by water are of two kinds: (a) material waste which contaminate the water; and (b) energy waste (for example, excess heat of thermal power plants) which raise the water temperature. When water returns to the environment carrying wastes, it influences it, very often in an undesirable way. Furthermore, the resultant changes in water quality will determine its further use for specific purposes, thus influencing water availability. Water quality is closely linked with the quantity of water available in a given region.

To be sure, natural water fulfills the transport function even without man's intervention. In its movement through the hydrological cycle, water carries with it dissolved materials of various kinds and suspended matter, inert and living, which may make it unsuitable for many purposes. Man's intervention in this process may be beneficial (for example, treatment of municipal water supply) or detrimental to the environment (for example, discharging waste into a stream). The central issue of managing the environment—and the quality of water resources that are part of it—appears to be that of striking the right balance between manipulations of natural systems necessary for the maintenance of desired socioeconomic structures and negative side effects produced by them.

The interest in the maintenance of certain levels of environmental quality is worldwide (Falkenmark, 1984). However, there are serious difficulties in reaching a consensus of what specific issues need to be addressed in the various regions of the world in order to reach the desired balance between the development and utilization of natural resources and the undesirable by-products of developmental activities, as witnessed by the United Nations Stockholm Conference on the Human Environment in 1972. It may not be much easier to reach such a consensus at national levels, but in the United States, scientists—within the government and outside it—have devoted considerable effort in the last decade in identifying and analyzing problems connected with water quality. It seems appropriate to determine now, the long-term research and development topics related to the protection of the nation's surface waters and groundwater resources requiring special attention by the government and by the private sector during the remainder of this century. This is the objective of this chapter.

One should clarify the term "protection" and place it in its proper context. Protection is the act of protecting something and sheltering it from danger or harm. Indeed, the surface waters and the groundwater resources have to be protected from dangerous and harmful substances and organisms. However, protection is only an element of *conservation,* which is defined as managing resources for sustained yield. Thus, the view taken in this chapter is that water resources—surface and groundwater—have to be managed so that their availability will be sustained beyond the current generation and the twentieth century. Their protection against contaminants will be a major contribution in this direction.

A key question underlying possible strategies for protecting the quality of the nation's water resources is the estimation of how much fresh water—whether in streams or in aquifers—is so contaminated that it cannot be used anymore.

The information regarding this loss of resource is sparse and sketchy. For example, about 3 million people were affected by well closings in New Jersey, Delaware, New York, and California, but we do not know the amount of water that became unfit for human consumption (Pye and Kelley, 1984). Five wells supplying drinking water in Tucson, Arizona, and seven wells in Phoenix were closed, but we do not know what is the volume of water lost (U.S. General Accounting Office, 1984). The Environmental Protection Agency (1984) estimates that 7,741 private, public, and industrial wells were closed or affected by contaminants without specifying the amount of water withdrawn from use. The U.S. Geological Survey is initiating a nationwide investigation regarding the degradation of water quality in streams and aquifers, including an assessment of the quantities lost for beneficial uses, and a report is contemplated in 1987.

In this study, surface waters are treated separately from groundwater. Although from the hydrologic point of view this is an aberration, most of the current scientific literature related to water quality is segregated along similar lines. The reason for it is not entirely clear, but it could be due to the fact that the separation between surface streams and groundwater resources is a long-standing tradition prevading institutions, laws, and research activities. We shall follow this tradition.

SURFACE WATERS

Overview

The quality of water bodies reflects in general the kind and extent of human activities in a region. The characteristic of the quality of water bodies in the United States is the presence of anthropogenic contaminants in most of them, in concentrations that vary from place to place. In many parts of the country, some drinking water sources have been seriously degraded by anthropogenic pollutants.

Systematic information gathering activity regarding water quality on a national basis was initiated by the United States Geological Survey in 1972. The National Stream-Quality Accounting Network (NASQAN) was established in that year, in order to (a) account for the quantity and quality of streamflow within the United States, (b) develop a large-scale indication of how stream water quality varies from place to place, and, (c) detect changes in stream quality with time. This monitoring activity is performed by means of a network of 504 stations. The data collected by the NASQAN network are point measurements; neither upstream nor downstream conditions are reported, so that there is little (if any) information regarding the variability of quality parameters along a stream reach. Moreover, neither viruses nor synthetic organic substances are systematically monitored to any appreciable extent (United States Geological Survey, 1984).

Recently, data covering the eight-year period 1974–1981 were analyzed

(Smith and Alexander, 1983) and the following trends in selected water quality parameters were detected:

- sodium and chloride concentrations *increased* in 34 percent of the NASQAN stations;
- sulfate concentration *increased* in 27 percent of stations;
- cadmium concentration indicated *no change* in 87 percent of stations;
- iron and manganese concentrations have shown *no change* in 84 percent of stations;
- alkalinity *decreased* in 26 percent of stations;
- calcium concentration *decreased* in 27 percent of stations;
- fecal streptococcus counts *decreased* in 29 percent of stations.

These findings seem to indicate a modest success in the efforts of cleaning up the streams of the nation. The decrease in fecal streptococcus bacterial counts indicates some improvement in the treatment of domestic wastes prior to their discharge into streams, and the essentially constant concentration of heavy metals reflects probably more extensive treatment of industrial wastes. The decrease in alkalinity may be due to the drop in pH of precipitation over large areas of the United States.

Drinking Water Issues

Average drinking water consumption in the United States is 2 liters per day. Undesirable constituents of drinking water can be natural or synthetic, entering surface waters by natural processes or by the activities of man. The contamination of drinking water sometimes can be easily perceived (by means of taste or color, for example), or can be detected only with special instruments and procedures. As a result, drinking water supplies are treated before their distribution to users. The treatment typically consists of the addition of chemicals which precipitate dissolved and suspended matter; then the water is disinfected so as to destroy microorganisms. The effectiveness of the treatment is less than 100 percent (sometimes, considerably less). Furthermore, the addition of disinfecting chemicals to drinking water may produce organic contaminants—for example, trihalomethanes—which may present a health risk even at low concentrations.

The water quality parameters in the United States are of two kinds: (1) National Interim Primary Drinking-Water Regulations (a list of maximum concentrations of 17 chemicals, turbidity, coliform bacteria, and three types of radionuclides), which are legally enforceable (U.S. Environmental Protection Agency, 1982); (2) National Secondary Drinking-Water Regulations (maximum levels of chlorides, sulfates, color, heavy metals, corrosivity, dissolved solids, and foaming agents), which essentially are guidelines for the states and are not federally enforceable. As of 1983, all states and territories assumed primacy, issuing drinking water regulations at least as stringent as the federal laws, with the exception of Indiana, Oregon, Pennsylvania, South Dakota, Wyoming, and the District of Columbia.

The main issues related to drinking water appear to be the following:

1. most synthetic contaminants are generally unregulated;
2. infrequent outbreaks of waterborne diseases still occur in this country. They are caused, in the majority of cases, by the presence of pathogens in the drinking water; and,
3. only water supply systems that have at least 15 service connections and regularly serve at least 25 individuals are subject to the Safe Drinking-Water Act of 1974.

The last point above refers to the fact that a large segment of the rural population is not protected by drinking water regulations. Indeed, an analysis of water samples from over 2,600 rural households in 1982 (U.S. Geological Survey, 1984) has shown that the concentration of total coliforms, lead, cadmium, and mercury exceeded the national interim primary drinking water standards in at least 15 percent of the cases.

Man's Interventions in the Geochemical Cycle

Anthropogenic Sources of Contaminants

The gross load of pollutants in the surface waters of a region depends on the following: (1) the density of population, (2) number and kind of livestock, (3) extent of industrial activity, (4) type of waste treatment, and (5) land use patterns. It should be mentioned that the total chemical flux in a stream is the sum of the waste discharges (anthropogenic), which are independent of the stream discharge, and of the background fluxes, which are proportional to the streamflow (Rickert et al., 1976).

Population is a significant contributor of organic carbon (40–50 g per capita per day), nitrogen (10–15 g/capita/day), and phosphorus (3–4 g/capita/day) (Sevier and Stumm, 1982).

Livestock produce contaminating agents which find their way to surface waters either when grazing on range land or when kept in feed lots. Surface runoff transports organic material and microorganisms from their excreta to streams and lakes.

Industrial by-products which have no use at the point of their production or at the time they are produced are usually termed *waste* and discharged into the environment. A significant portion of them reach surface waters, degrading their quality.

Wastewater treatment methods appear to be based primarily on the principle of separating the most offending contaminant and dealing with it subsequently in a manner that is not always clear. This principle may be effective in the case of conservative pollutants (such as heavy metals), but it does not seem to be so in the case of organic material.

Land use patterns may affect the kind and concentration of contaminants. For example, agricultural lands will contribute primarily fertilizers and pesticides to the chemical load of surface waters, while urban areas will tend to discharge synthetic organics and heavy metals.

Chemicals of Anthropogenic Origin

Nearly half of the world output of chemical industry, estimated at 100–200 million tons per year (Sevier and Stumm, 1982), is generated in the United States. The United States chemical industry produces about 4 million known substances, of which approximately 2,000 are produced at a rate greater than 500 tons per year. Many chemicals are distributed throughout the country and some chemical products are among the most dangerous contaminants of surface waters and most recalcitrant regarding their removal and disposal.

Synthetic organic chemicals are all-pervasive. We use them in foods (as additives, preservatives, artificial flavoring, coloring and enrichment agents), as cleaning agents, detergents, solvents, pesticides, and fertilizers. Many of them are known not to be biodegradable and may be harmful to man, either directly or through biomagnification as a result of their concentration in other organisms.

In addition to exhibiting health risks to man, animals, and aquatic organisms, chemicals of anthropogenic origin may also interfere with natural food chains in the environment, both in surface streams and in groundwater.

Point and Nonpoint Sources of Pollution

Contaminants may enter surface waters (and for that matter also groundwater aquifers) either at distinct points, or over extended areas. Point sources of pollution are mostly those associated with discharges of liquid municipal and industrial wastes, treated or untreated. Nonpoint sources are all the other.

The most affected streams in the United States are those in the Northeast and Mid-Atlantic regions where over 50 percent of the United States population lives and works (U.S. Geological Survey, 1984). In these regions, one finds six of the ten leading chemical-producing states, accounting for 41 percent of the nation's chemical production. These chemical industries are significant point sources of contaminants in the northeastern United States.

In the same section of the United States, nonpoint sources of pollution are also of importance. The salt used for road deicing to the extent of about 9 million tons per year (Sevier and Stumm, 1982)—100 tons per km^2 in the Boston area alone—eventually finds its way to surface streams.

Man's activities result also in the acceleration of the nutrient cycle in nature, leading to the enrichment of surface waters with phosphorus and nitrogen through point sources. Added to the point sources of pollution are the discharges of nonpoint sources of contamination which commonly contain high concentrations of nutrients, herbicides and insecticides, heavy metals, petroleum products, and oxygen-demanding materials. In urban areas, pollution from nonpoint sources commonly equals or exceeds that from point sources of wastewater.

Urban runoff is a major nonpoint source of heavy metals entering surface waters. Compared with shale—a sedimentary rock considered to represent the composition of sediments deposited in the absence of man's influences—street sweepings are considerably richer in lead (by a factor greater than 90), cadmium (more than 11 times), zinc (more than four times richer), the concentration of some of the heavy metals exceeding significantly the acceptance criteria for public water supplies (U.S. Geological Survey, 1984).

Agricultural pesticides and insecticides are important contaminants from nonpoint sources and continue to appear in surface waters, albeit in lower concentrations, although many of them were banned in 1975. Some of them, such as DDT and other organochlorines, sorb strongly to soil and enter surface waters as a result of soil erosion and sediment deposition. Due to their resistance to decay, it is expected that their appearance in surface waters will continue over a considerable period of time.

No discussion of nonpoint sources of pollution would be complete without referring to the salinity of surface waters. A comprehensive and clear picture covering the entire country is not available. However, significant information exists regarding the Colorado River basin. In this river basin, total dissolved solids (TDS) vary from 50 parts per million (ppm) in the headwaters of Colorado to about 800 ppm of TDS at the Imperial Dam (U.S. Bureau of Reclamation, 1983). The increase in salinity is attributed as follows:

- From natural sources—47 percent
- Due to irrigation—37 percent
- Due to loss of low salinity water by evaporation from reservoirs, transbasin diversions, and municipal and industrial consumption—16 percent

Economic losses due to salinity in the Colorado River basin were estimated in 1982 to amount to $113 million.

Acid Precipitation

Industrial and agricultural activities release into the atmosphere a great variety of pollutants. Following their dispersion in the atmosphere, many of these contaminants (including heavy metals) are carried by precipitation back to the land surface and then appear in the hydrosphere (Lantzy and McKenzie, 1979).

Serious injuries to the aquatic environment via the atmosphere are inflicted by acid rain. Increased burning of fossil fuels increases locally the concentration of carbon dioxide and other gaseous oxides, lowering the pH of rain. Increased acidity of precipitation results in higher weathering rates of the lithosphere, thus releasing in solution a host of ions which include phosphates, aluminum and trace metals.

To be sure, the atmosphere also contains alkaline material, such as soil particles, which tend to neutralize the effect of the carbon dioxide. However, in the northeastern part of the United States, the large concentration of industries burning fossil fuels upsets this balance and produces acid rain.

Precipitation, unaffected by industrial emissions, has a pH of about 5.0. The precipitation in the northeastern United States has a pH of 4.8 to 4.2. In addition, acid rain with a pH less than 4.8 was observed also in eastern Arizona, western New Mexico, and southwestern Colorado (Interagency Task Force on Acid Precipitation, 1983). The reason seems to be the ore-smelting activities in this mineral rich region.

Ultimately, when acidic rain enters water supplies, it can corrode copper and lead pipes and, if used for public water supplies, special treatment is required to protect the public health from the toxic effects of these metals.

Streams, Lakes, Estuaries

Eutrophication

Eutrophication is the process by which surface waters become enriched with plant nutrients, mainly phosphorus and nitrogen, leading to increased production of algae and aquatic weeds. It is one of the most important problems in industrial societies, where phosphate is the critical component determining productivity of the biomass (Schindler, 1977).

Point sources contribute mostly toward the phosphorus loading of the receiving water bodies, as was the case in Lake Erie about a decade ago (Flynn, 1982). Now that the use of phosphates in household detergents is somewhat controlled, the phosphorous loading from anthropogenic wastes of Lake Erie has been reduced (U.S. Geological Survey, 1984).

The nonpoint sources, primarily the contribution of agricultural lands, present quite a different aspect. The main contributors to the eutrophication process are nitrogen and phosphorous fertilizers. Almost all nitrogen fertilizers (nitrates, ammonia salts, urea) are highly soluble and most of the portion which is not utilized directly by the crop to which they were applied are transported by rainfall and return irrigation flow to the nearest surface water bodies. To be sure, a significant proportion may also be leached through the soil profile to the local groundwater table.

Phosphates, by contrast, are much less soluble than nitrogenous fertilizers and furthermore are immobilized by the clay fraction of many soils. In order to ensure effective phosphorus fertilization, farmers use somewhat heavier applications of phosphorous fertilizers in order to increase their effect on the crop grown. It follows, then, that substantial quantities of phosphate are immobilized in the agricultural lands of the United States, only to be released slowly in soluble form. Thus, even more judicious use of fertilizers in agricultural production will not remove completely the eutrophication problem, although it could certainly control it to a large extent.

Eutrophication can occur in all surface water bodies: streams, lakes, and estuaries. It is more acute, however, in stagnant waters, where hydrodynamic dispersion and dilution are at the minimum.

Contamination by Bottom Sediments

Bottom sediments tend to concentrate a broad variety of pollutants, from organics—inert and living—to heavy metals. These pollutants may present health risks to man, animals, and aquatic organisms, as well as interfering with food chains in nature.

Contaminated sediments may act as a source of pollution for considerable periods of time, long after the discharge of the source material ceases. Some heavy metals, such as mercury, may remain a hazard for many years, possibly for centuries (U.S. Geological Survey, 1984), because no quick solution for their treatment and reclamations seems readily available.

Recommendations

It seems that in order to continue the maintenance of the environmental quality and to increase the degree of protection of the nation's surface water, long-term studies and research are needed in two directions:

(1) wider and more detailed monitoring of the quality parameters of surface waters, particularly with respect to sources of contaminating agents and their destination in surface water bodies; and
(2) broader and more precise knowledge of natural processes affecting the quality of waters, especially hydrogeochemical and hydrogeobiological phenomena.

More specifically, the following is a partial list of suggested studies, not necessarily ranked in a priority order:

Augmentation of NASQAN activities. The surface water quality monitoring network takes currently point measurements. The information thus acquired relates to discrete points of phenomena which are probably best represented by curves in multidimensional space. It is necessary, therefore, to augment the NASQAN measurements to yield, in addition to point estimates of water quality parameters, approximate values of the directional first derivative of the dynamic changes of these parameters. In this way, one could have a better idea of the contamination-purification processes in the time-space continuum. The NASQAN network also should be augmented to include the monitoring of synthetic organics, in addition to the contaminants monitored currently, and of viral organisms, in addition to the fecal coliforms and fecal streptococci. In this way, one could reach a quantitative assessment of the nation's surface waters with respect to their capacities to absorb and purify contamination.

Surface water diversions. Mineral content of surface streams usually increases from source to outlet. Diverting water from a stream will tend to deprive downstream users of higher quality water. The proposed study will have a double objective:

1. determine the increase in salinity of surface waters due to existing diversions, leading to improved operating policies for water diversions in order to minimize undesirable salinity effects; and
2. establish criteria for future diversions from surface streams which would (a) minimize downstream increases in salinity, and (b) apportion costs of increased salinity in an equitable manner.

Pathogens in drinking water. Epidemiological studies have shown significant correlations between disease and origins of potable water sources (Cantor and McCabe, 1978). Because at least 50 percent of the U.S. population drinks water from surface sources, it is important to broaden the studies of pathogenic substances in drinking water to include a broad range of viral organisms and genetic material, some of which is suspected to be carcinogenic. Under the same heading, one should include studies of drinking water disinfection and the development of new alternative methods. The current chlorination procedures seem to be considerably less than effective, since excess chlorine can combine with organic material to form chloramines, which are probably carcinogenic.

Accumulation of contaminants in food chains. It has been observed that DDT accumulates in certain fish, and viruses accumulate in oysters. Any substance obtained from surface waters—fresh, brackish, or saline—used for human consumption or as a food additive (for example, agar extracted from seaweeds) has the potential of transmitting hazardous matter to man. It is necessary, therefore, to determine the tendencies of various contaminants, inert as well as living, to accumulate at various links in the natural food chain and to become biomagnified at the points of accumulation. This study should yield improved criteria for treating wastes and their discharge into surface water bodies.

Wastewater treatment. Current methods of treating municipal and industrial wastewaters is by attempting to *separate* substances considered noxious from the water itself. In this way, one attempts to concentrate, as far as possible, the contaminants, releasing the treated water back to nature or for some specific use, then disposing of the sludge in some fashion. In many cases, the significant efforts invested in the separation procedure have only limited effects in upgrading the quality of the wastewater, yet essentially they transfer the problem of dealing with contaminants from the aquatic environment to the dry land. This issue will appear later in this study when landfills will be discussed in connection with protecting groundwater resources. What appears to be needed is a novel approach to wastewater treatment based on the principle of *transforming* the contaminating organic material from its existing form into harmless biomass. Laboratory experiments have demonstrated the feasibility of biodynamic treatment of wastewater, by which raw domestic sewage was acted upon by a succession of organisms resulting in reclaimed water, low in nitrates, with practically no sludge, and in which healthy fish were swimming. This type of research and development is needed.

Biodegradability of synthetic organic compounds. The chemical industry is extremely prolific in synthesizing, almost daily, new organic compounds. The industry is furthermore imaginative enough to transform most of these synthetic organics into marketable products. Alongside these admirable qualities, the chemical industry should also recognize its responsibility toward the public-at-large and not only to its current shareholders. One step in this direction would be the determination of the biodegradability of existing and each new synthetic organic compound. At the very least, the public will know that it is faced with a new, nonbiodegradable product and measures could be devised for handling such wastes in appropriate ways. At the most, such products could be banned altogether and only biodegradable and nonbiodegradable compounds that do not present a hazard to the environment would be allowed to be marketed. Between these two extremes, a range of alternatives may be developed.

Studies of hydrogeochemical and hydrogeobiological processes. Two examples will be offered under this group of studies: (a) the dynamics of phosphate solubility need to be understood better, including the hydrogeochemical reactions and the biological factors contributing to the release of phosphates in solution; (b) the accumulation of contaminants in bottom sediments in rivers, lakes, and estuaries, and subsequent release through biochemical reaction in natural food chains.

GROUNDWATER RESOURCES

Overview

One of the major issues facing us today is the growing concern over groundwater contamination. It is not visible, it is insidious, and its negative effects become obvious in the form of crises. It is quite apparent that groundwater contamination will be a major problem for society during the foreseeable future.

Contaminants of groundwater—as is the case for surface waters—may be organic or inorganic in nature. The hydrogeochemistry of inorganic constituents is better understood than that of the organic compounds. The organic substances, however, pose more serious problems: they may remain, in some places, as a source of groundwater contamination for long periods of time (Bredehoeft, 1983).

The importance of groundwater resides in the fact that about 75 percent of U.S. cities derive their water supplies in total or in part from groundwater (Pye et al., 1983). The use of groundwater grew spectacularly since 1950. From an average use of 34 billion gallons per day (bgd) in that year, groundwater supplies doubled to 68 bgd in 1970, only to grow yet an additional 30 percent to 88.5 bgd in 1980. Most of groundwater (68 percent) is used for irrigation.

It is estimated that total volume of groundwater within 2,500 ft. (750 m) of ground surface amounts to between 33 and 100 quadrillion gallons. It is also estimated that between one and two percent of groundwater near land surface is contaminated (Pye et al., 1983).

Today's challenges related to the protection of groundwater resources are of three kinds:

(1) Prevention of the introduction of contaminants into aquifers, especially since chemicals of all sorts are part of our contemporary culture;
(2) Predict the behavior of the contaminants once they have entered the aquifer systems;
(3) Remove the contaminants from the aquifers so as to protect effectively the biosphere (U.S. National Academy of Sciences Geophysics Study Committee, 1984).

Aquifers may also have an opposite effect on contaminants. The toxicity of the pollutants may be attenuated through dilution, volatilization, mechanical filtration, precipitation, buffering, neutralization, and ion exchange (Pye and Kelley, 1984).

Sources and Mechanisms of Contamination

Groundwater may be contaminated by natural processes or its contamination may be due to man's activities. The natural processes most commonly observed are leaching of solutes through the soil profile and the vadose zone, causing the mineralization of groundwater, and evapotranspiration which tends to concentrate the dissolved solids in the percolating waters (Matthes, 1982).

The anthropogenic sources of groundwater contamination may be classified into two categories: (1) those due to waste disposal practices, which can also be considered as point sources; and (2) those unrelated to waste disposal practices, which are essentially nonpoint sources of pollution.

Among the most important point sources of anthropogenic contamination of groundwaters are sewage disposal systems. Each individual system, through its lagoons, ponds, and other components, is a potential contributor of pollutants to the local water table. In addition, the entire network of pipes and sewers collecting the wastewaters and transporting them to the treatment plant may be a source of contamination. Septic tanks are serious sources of pollution in areas not serviced by sewerage.

Land disposal of solid wastes by means of landfills and land spreading of sludges present serious hazards to the quality of groundwater resources. Contaminated liquids may seep from these sites, eventually percolating to the local and regional water tables.

Evaporation ponds and lagoons used by various industries, including those for the disposal of brines from oil fields and disposal of mining wastes, may leak hazardous contaminants to the water table. Even if originally lined, such ponds and lagoons will eventually become a source of contamination due primarily to faulty maintenance and neglect to repair or replacement of defective lining.

Deep well disposal of liquid wastes and burial to depths in excess of 300 m of hazardous and radioactive wastes seem attractive techniques related to waste disposal practices. However, these practices present serious hazards, primarily for two reasons. First, most deep injection wells are located in dense rocks which, nevertheless, exhibit networks of fine fractures; very little is known regarding the hydrologic properties of fractured rocks. Second, shafts and boreholes leading to deep repositories can act as a short circuit for the back flow of wastes to shallow levels and thus contaminate exploitable aquifers.

Among the anthropogenic sources of groundwater contamination are the following: accidental spills and leaks of chemicals and wastes; agricultural activities involving fertilizers and pesticides; mining activities, especially disposal of mine tailings; highway deicing by the use of sodium chloride; atmospheric contaminants introduced into the hydrological cycle by acid rain; contaminated surface waters hydrologically connected with aquifers; saltwater intrusion, primarily due to faulty planning of groundwater development and utilization; and defective well construction and improper maintenance.

The physical mechanism by which aquifers become polluted is the transport of contaminants. From a chemical point of view, transport phenomena are of two kinds: conservative transport, and transport involving chemical reaction.

Conservative transport in aquifers occurs through hydrodynamic dispersion which is due to variations in the permeability of the water-bearing material. Although for most analyses the hydraulic conductivity of an aquifer is assumed to be constant in time and in space, it is neither. Permeability variations within an aquifer can exhibit a range of over fifteen orders of magnitude (U.S. National Academy of Sciences Geophysics Study Committee, 1984).

Transport of contaminants involving chemical reaction is analyzed mostly in either of two ways. One approach assumes that chemical equilibrium is rapidly attained and the other takes into account the reaction kinetics. In many cases, some constituents of wastes may interact chemically forming new and unexpected compounds. These phenomena, together with the microbial activity in the soil profiles through which contaminated water percolates, makes it difficult to predict what pollutants eventually reach an aquifer.

Sources of groundwater pollution vary throughout the United States (U.S. Congressional Research Service, 1980). On a regional basis, the major sources of groundwater contamination are as follows (Pye and Kelley, 1984):

- Northeast (New England, New York, New Jersey, Pennsylvania, Maryland, Delaware): septic tanks, landfills, highway deicing, abandoned oil wells.
- Northwest (Colorado, Idaho, Montana, Oregon, Washington, Wyoming): dryland farming, irrigation return flow, septic tanks, buried pipelines and storage tanks.
- Southeast (Alabama, Florida, Georgia, Mississippi, North Carolina, South Carolina): underground storage tanks, landfills, surface impoundments, accidental spills.
- Southcentral (Arkansas, Louisiana, New Mexico, Oklahoma, Texas): mineralization from the soluble components of the aquifer matrix, over-pumping, disposal of oil field brines.
- Remainder of States (Great Lakes, North Central, Alaska, Hawaii): not known.

Severity of Contamination

The U.S. Environmental Protection Agency estimated that between 0.1 and 0.4 percent of exploitable aquifers are contaminated by industrial impoundments and landfills; about 1 percent are polluted by septic tanks; and petroleum exploration and mining activities contribute to the contamination of further 0.1 percent (Pye and Kelley, 1984). This gives us an idea of the *extent* of groundwater contamination in the United States.

The *severity* of contamination could be expressed quantitatively by the following attributes of the contaminants (Pye and Kelley, 1984): concentration, persistence, toxicity, number of persons affected if the aquifer is a source of drinking water, percentages of available groundwater affected, and the economic cost of alternative water supplies. There are serious doubts regarding the adequacy of expressing the severity of contamination in this way primarily for two reasons. First, almost nothing is known regarding prolonged exposure to low concentration of pollutants. We have virtually no idea about the physiological or genetic effects of drinking water having minute concentrations of heavy metals during several decades. Second, concentration seems to be an inadequate yardstick for measuring severity of contamination by viruses; the presence of one virus in drinking water appears to be a public health hazard. Furthermore, we do not know anything in relation to possible synergistic effects of both contaminating chemicals and microorganisms. This appears to be one of the major problems to be elucidated regarding the protection of the nation's waters.

Major Contaminants

Saline Water Intrusion

Sea water intrusion is of particular interest in coastal aquifers (affecting groundwater) and in estuaries (affecting surface streams). Saline waters originating in oil fields and brackish aquifers may also intrude into regional groundwater resources.

In coastal aquifers, under conditions of equilibrium undisturbed by man's activities, an "interface" can be detected which separates the freshwater from seawater. Deep wells or wells too close to the shoreline disturb the equilibrium represented by the interface between two bodies of water with different specific gravities. Because of this difference, depressing the water table by 1 foot will produce an upconing of the interface of about 38 feet. As a result, the wells may start to draw a mixture of fresh and seawater, the salinity of the pumped water increasing with time. Thus, extensive pumpage from a coastal aquifer reduces the seaward discharge of fresh groundwater and causes the interface to shift upward and landward.

Following development and exploitation of coastal aquifers, the fresh water-saline water interface may be stabilized at a predetermined position further inland and a new equilibrium can be established. In this equilibrium, groundwater pumping is balanced by a reduction of the natural discharge of fresh water into the sea, and by an increase in the rate of recharge of the aquifer.

Several strategies can be employed for stabilizing the interface in coastal aquifers and thus protect the groundwater from sea water intrusion. One such strategy consists of injecting surface water or treated wastewater into a series of wells roughly parallel to the coastline in order to create a "fresh water barrier" which would retard the inland advance of the interface. Such barriers are quite successful in protecting the Los Angeles groundwater basin in Southern California.

Synthetic Organic Contaminants

Although the contaminants most commonly reported in literature are chlorides, nitrates, heavy metals, and hydrocarbons, reflecting probably the most accepted monitoring practices related to groundwater quality, contamination by synthetic organic chemicals is significant and probably even more serious than what was envisioned a decade ago (Pye and Kelley, 1984).

The magnitude of this contamination is uncertain, but it is estimated that between 0.5 and 2 percent of the groundwater within 200 feet of land surface may be contaminated by these compounds (Trussell et al., 1984). The seriousness of this contamination can be appreciated considering that over 50 percent of the U.S. population is supplied with water from aquifers. Synthetic organic contaminants seem to be persistent in groundwater, so that the contamination of the aquifers could be irreversible (U.S. Council on Environmental Quality, 1981). To make matters more complicated, some of the synthetic organic contaminants are volatile. Whereas when discharged in surface waters they may be largely removed by natural aeration, their presence below the land surface seems to be considerably more protracted. A very complex dynamic equilibrium is established between the gaseous phase of the contaminant in the vadose zone and the liquid phase in the saturated zone of the aquifer. Although some of the contaminant may escape to the atmosphere in vapor form, its complete volatization can be a process of considerable duration.

A major problem related to synthetic organic contaminants is the lack of competent analysts and adequate equipment in government laboratories for the identification of these pollutants. The art and science of analytical chemistry apparently did not keep pace with the proliferation of synthetic compounds produced by the chemical industry and distributed into the environment.

Related to the problem is the issue of monitoring. Information gathering regarding groundwater quality should be broadened to include at least some of the more common synthetic organic compounds.

Hazardous Wastes

Hazardous waste is any substance that is toxic or otherwise a threat to life when discharged through human activity to the land, water, or atmosphere (U.S. Geological Survey, 1984). Groundwater contamination from hazardous wastes generated a great deal of interest lately, including mathematical modeling of the phenomena identified, and the numerical simulation of the processes involved (Wood et al., 1984).

Groundwater resources are contaminated by hazardous waste through faulty waste disposal methods. The hazardous wastes may reach the groundwater through landfills, dumps, and buried wastes, or by means of contaminated liquids through pondage, direct injection in waste disposal wells, and spills and leakage (U.S. Environmental Protection Agency, 1980).

During their slow movement from ground surface to the water table, hazardous wastes may be attenuated through dilution, dispersion, sorption, ion exchange, precipitation, and other physicochemical and biochemical reactions. Contamination is thus limited to short distances around the point source, but may become serious in densely populated areas with many waste disposal sites in close proximity to each other.

Contamination of groundwater by hazardous wastes became serious enough that in 1980 Congress passed the Comprehensive Environmental Response, Compensation, and Liability Act for the purpose of identification and remedial cleanup of abandoned hazardous waste sites. Also this legislation authorized the funds necessary for these operations, which are referred to as "Superfund." So far, the U.S. Environmental Protection Agency (USEPA) has identified about 14,000 such sites (U.S. Geological Survey, 1984). Current remedial measures used in areas where groundwater is contaminated by hazardous waste include interceptor wells, physical barriers, and relocation of public supply wells. A methodology for ranking the relative risk of groundwater contamination from hazardous material sites was recently proposed (Olivieri and Eisenberg, 1984).

Radioactive Wastes

Radiation above the natural average background is considered a hazard to public health. Low level wastes generally average less than one curie of activity per cubic foot of material, or less than 10 nanocuries of transuranic contamination per gram (Lipschutz, 1980). Low level radioactive waste is generated mostly at hospitals, research facilities, and nuclear power production plants. It amounts annually to about 210,000 cubic yards (U.S. Geological Survey, 1984). This type of radioactive waste is usually disposed of by burial in shallow trenches at commercial and federal sites. Several of these sites exhibited geohydrologic problems and were closed, or their closure is under consideration in courts. In short, the safe disposal of low level radioactive wastes has at best been only a partial solution.

High level wastes are officially defined as the waste streams that result from the reprocessing of spent reactor fuel. These wastes, which are characterized by nuclides with relatively long half-lives and emission of considerable amount of heat, have no suitable permanent repository yet. The current approach under consideration is to isolate these wastes from the biosphere by means of multiple barriers relatively independent of each other that would impede the migration of radionuclides away from the repository. Migrating radionuclides were observed to have moved about 2 km away from repository and subsequently were adsorbed and remained immobile by the geological formations through which they traveled (U.S. National Academy of Sciences Geophysics Study Committee, 1984). It follows that the establishment of permanent repositories for high level radioactive waste should take into account the possible migration of radionuclides and their adsorption by the solid matrix of aquifers, thus becoming a long-term contamination source of groundwater.

Because of their composition which includes nuclides with long half-lives, high level radioactive wastes have to be isolated from the biosphere for periods of time measured in thousands of years (U.S. Nuclear Regulatory Commission, 1983). The long-term containment of these wastes should consider also the possibility of significant climatic changes such as occurred in the Pleistocene, when glacial periods alternated with the retreat of glaciers.

Legal Framework

The extent of groundwater contamination in the United States is not known, yet it is considered to be a serious national problem primarily due to hazardous waste. Major contamination problems exist in all states east of the Mississippi where a large part of the industry is located, as well as in several nonindustrialized states west of the Mississippi, such as Arizona and Idaho (U.S. General Accounting Office, 1984).

The need to protect groundwater quality by a legal structure was recognized as a major national concern by the EPA (U.S. Environmental Protection Agency, 1984). Yet there are no specific laws oriented toward groundwater (with the exception, perhaps, of the Arizona Groundwater Act of 1980). Worse, almost all statutes seem to ignore totally the hydrological connection between surface streams and groundwaters and the link between land use and groundwater quality (Pye et al., 1983).

Although no federal legislation is directed specifically to comprehensive groundwater protection, the following laws address specific sources of aquifer contamination:

- Clean Water Act, 1972, focusing on surface water pollution,
- Safe Drinking Water Act, 1974, oriented toward making sure that water supplied to the public is safe to drink;
- Resource Conservation and Recovery Act, 1976, directed toward the safe disposal of hazardous wastes and the management of these disposal activities;
- Comprehensive Environmental Response, Compensation, and Liability Act, 1980, providing for emergency response and clean-up of hazardous waste contamination ("Superfund" Act);
- Surface Mining Control and Reclamation Act, directed toward protecting the environment from adverse effects of strip mining of coal;
- Uranium Mill Tailings Radiation Control Act, concerned with the effect of uranium and thorium wastes on the environment, including surface streams and groundwater.

Recently, EPA issued a proposed strategy for protecting the nation's groundwater (U.S. Environmental Protection Agency, 1984). Two main issues were identified in this proposal: (a) protection of the public health; and (b) protection of critical environmental systems. With these objectives in mind, aquifer systems were classified into three categories:

1. special aquifers, which are the sole source of drinking water and are irreplaceable;
2. aquifers which are current or potential sources of drinking water;
3. all other aquifers, some of which may be saline, or contaminated beyond reclamation and use.

The EPA policy draft paper recognizes three sources of groundwater contamination: (1) waste disposal activities; (2) use of chemicals on land; (3) overpumping of aquifers. Most emphasis is placed on the waste disposal activities.

EPA estimates that there are about 16,000 potentially hazardous waste disposal sites, of which 5,600 (or 35 percent) have been screened and 539 were listed for priority action under the Superfund legislation. In addition, there are in the United States about 93,500 landfills, of which approximately 75,000 (or 80 percent) are industrial dumps and almost nothing is known about them. The remainder 18,500 landfills are managed by municipalities. There are also about 181,000 impoundments of liquid industrial wastes (such as mine tailing ponds), most of which are not lined. EPA estimated that only about 7 percent are located and constructed so as to pose little or no threat to groundwater. Finally, septic tanks are used by about 20 million households in the United States, each handling a flow of about 50 gallons per day. The aggregated discharge amounts to approximately 3.5 bgd—a sizeable flow. In addition, there are commercial and industrial septic tanks about which almost no information exists.

This rough sketch of the current status of groundwater contamination in the United States seems to justify a clear policy for managing the quality of the nation's groundwater resources.

Recommendations

In order to protect effectively the nation's groundwater resources, it is necessary to have a clearer and more complete understanding of the *mechanics* of groundwater contamination and of its *consequences* insofar as public health and environmental quality are concerned. The following studies, listed in no particular order of priority, will contribute toward reaching that level of understanding.

Transport phenomena in the vadose zone

The passage of contaminated waters through the vadose zone triggers chemical reactions and transport phenomena that are not very well known. There is evidence that many of the chemical reactions involving organic compounds in the vadose as well as the saturated zones of an aquifer are controlled by microorganisms (Wilson and McNabb, 1983). The entire problem of transport with the attending chemical and biochemical reactions needs elucidation. A corollary of this study would be the determination of effects of chemical reactions on transport and dispersion of contaminants.

Studies related to the disposal of radioactive wastes

Low-permeability rocks are currently considered to be good candidates for repositories of radioactive wastes. Such sites can effectively isolate toxic and radioactive wastes from the biosphere for very long periods of time. However, many rocks of low permeability exhibit networks of fine fractures oriented in many directions. The hydrological properties of fractured rocks is a new chapter in the science of hydrology. The exploratory research that was initiated in recent years has to continue and expand so that the hydrology of fractured rocks is fully understood.

Further development and refinement of micro-analytical techniques

Analytical techniques are already available for the determination of minute concentration (parts per billion) of certain contaminants. The range of these techniques needs to be expanded in two directions: (a) increased numbers and kinds of organic compounds which can be identified; (b) possibility of detecting micro-concentrations of pollutants, at the level of parts per trillion.

A comprehensive survey of groundwater contamination

What is really needed is a detailed survey of the nation's groundwater resources. The quality aspect of this survey would include descriptions of the severity of contamination wherever degraded quality of groundwater is encountered. Severity of contamination needs to be redefined, taking into account microbial and viral contaminating episodes. In this survey aquifers have to be classified from a physical (rather than utilitarian) point of view, into confined and phreatic aquifers. This classification is probably more relevant to managing groundwater quality, because in the case of confined aquifers protective measures have to concentrate primarily in recharge areas, while in the case of phreatic aquifers protective measures should be included in land use plans of areas overlying them.

Restoration of contaminated aquifers

New approaches are needed in the restoration of aquifers contaminated by hazardous wastes. Whereas current practices rely primarily on chemical and mechanical means, it is suggested that a vigorous research program should be initiated focused on microbiological activities. Bacteria and other microorganisms capable of breaking down organic contaminants should be identified, and their optimal conditions for growth determined. Then, by stimulating their development in contaminated portions of aquifers, hazardous wastes could be decomposed and more readily neutralized or removed.

Research related to the consequences of contamination

Short- and long-term tests need to be carried out for each substance suspected to be a hazard to public health or to the environment, in order to determine (a) synergistic effects in combination with other substances, (b) antagonistic effects in relation to other contaminating agents, and (c) effects of long-term exposure to low concentrations of the contaminant. Results of these investigations should enable governmental agencies to establish maximum limits on concentrations of various chemical contaminants in groundwater, and to update periodically these limits in the light of new findings. This should be an ongoing program, expanding in scope to include new synthetic organic compounds, as they are distributed for widespread use.

Land disposal of wastes

Methods of segregating, treating, and disposing of wastes on land should be reviewed and evaluated with regard to potential negative effects on underlying groundwater aquifers. The critical evaluation should identify gaps in knowledge and understanding of natural processes, thus opening new avenues of basic and applied research.

Ocean disposal of wastewater

Although disposal of wastewater in the ocean does not relate directly to the protection of groundwater resources, it is clearly an alternative for the alleviation of the contaminant load on aquifers. Ocean disposal of wastewater is carried out primarily through direct discharge, although the sinking of containers considered to be impervious and filled with liquid waste is also a possible option. The goal of the design of the wastewater disposal system discharging directly into the ocean is to minimize the detrimental effects of the discharge on the environment. This is accomplished primarily by rapid initial mixing of effluent with the ambient water. The subsequent turbulent diffusion and transport processes are dependent primarily on the flow pattern of the ocean water and hence controlled by nature (Fischer et al., 1979). Deeply submerged containers with liquid wastes may develop leaks and a plume of contaminant may appear. Its behavior will be determined by factors related to the ocean environment such as density stratification, temperature variations, and flow patterns (currents). Since these factors vary considerably from site to site and also from time to time at the same site, a good knowledge of the ocean environment is necessary when considering a certain site as a candidate for ocean disposal of wastewater.

SUMMARY AND CONCLUSIONS

The rapid expansion of industrial activities in the United States during the post–World War II years, the significant rise in the standard of living, and the spectacular growth of the chemical industry is resulting in the discharge of massive quantities of waste products into the environment. Within about a decade, the contamination of the environment (and in particular of the surface waters and of groundwater resources) reached in some areas critical proportions and caused considerable public concern. It took approximately another decade for this public concern to be transformed into federal legislation and state laws oriented to the protection of the nation's water resources but focused primarily on surface water bodies. The enforcement of these statutes has had positive effects—perhaps not spectacular—in the sense that many streams and lakes which in the sixties were polluted to the extent that very few living organisms could survive in them, today are considerably cleaner and have reverted to safe use by man. Yet another decade had to pass before the need to protect groundwater quality was recognized as an issue of major national concern.

Aquifers supply about one-quarter of the water used in the United States. In 1980, total water use averaged 372 bgd, of which 89 (or 24 percent) were pumped from the ground. Most of the groundwater (60 bgd) was used for irrigation, 17 bgd (or 19 percent) was supplied to industrial and other uses, and only 12 bgd (or 14 percent) was public drinking water.

Nevertheless, about 117 million people (or 50 percent of the population) gets their drinking water from aquifers, supplied by 48,000 community public water systems and 12 million wells. The other half of the U.S. population obtains its drinking water from 11,000 public water systems supplied by surface water. About 95 percent of rural households depend entirely on groundwater, as do also 97 percent of the 165,000 non-community public water systems (for example, restaurants on highways). Of the 100 largest U.S. cities, 34 rely entirely or in part on groundwater (U.S. Environmental Protection Agency, 1984). One of the greatest damages facing the nation's water resources is the presence of over 61,000 chemicals in the market place—most of them synthetic organics—their number increasing by several hundred each year.

One could argue that the issue of groundwater quality is overrated, since the U.S. Geological Survey estimated that the nation's aquifers contain a volume of water (mostly of high quality) sixteen times that of the Great Lakes, indeed an impressive amount. However, since so many people depend on aquifers for their drinking water, local contamination of these resources is a problem of considerable concern.

Management of groundwater quality—including its protection—can be effective only if we understand the processes governing the concentration of solutes during their percolation below ground surface. Some of the phenomena that ought to be studied are mentioned in another section of this study and need not be repeated here. However, it may be worthwhile to emphasize one aspect of the percolation phenomenon so as to dispel a misconception held by some.

Protection of surface waters and groundwater resources entails policies at the federal, state, and local levels. A policy can only be as effective as the understanding of the natural and social phenomena with which it deals. Policies not based on this understanding create more problems than they solve. The artificial separation between surface and groundwaters, both being parts of the same cycle in nature, is a case in point. Indeed, most states believe that one of their highest priority needs in relation to the protection of groundwater is a program of research for better understanding of hydrological processes (U.S. General Accounting Office, 1984). Policies should emphasize prevention of contamination rather than curative and reclamation efforts; the former is much less costly than the latter. Ongoing research should be the basis for updating policy.

A broadbrush sketch of some current problems related to the quality of the nation's surface waters and groundwater resources was presented. The current status of the scientific and technological base relevant to the protection of the U.S. fresh waters from contamination was described in each of the two sections devoted respectively to surface waters and groundwater resources. Each of those sections concluded with recommendations regarding long-term environmental research and development activities needed to fill some of the identified gaps in knowledge so as to solve some of the existing problems. The resolution of these problems is considered important for the continued maintenance of environmental quality and human welfare.

BIBLIOGRAPHY

Bredehoeft, John D. "Groundwater: A Review," in *U.S. National Report to International Union of Geodesy and Geophysics 1979–1982*. Washington, DC: American Geophysical Union, 1983, pp. 760–765.

Cantor, K. B. and L. J. McCabe. "The Epidemiological Approach to the Evaluation of Organics in Drinking Water," in *Water Chlorination, Environmental Impacts and Health Effects,* R. L. Jolley et al., editors. Ann Arbor, Michigan, 1978, pp. 379–394.

Falkenmark, Malin. "Water—The Silent Messenger Between Cause and Effect in Environmental Problems." *Water International* 9, 2 (1984):62–65.

Fischer, Hugo B., E. John List, Robert C. Y. Koh, Jörg Imberger, and Norman H. Brooks. *Mixing in Inland and Coastal Waters*. New York: Academic Press, 1979

Flynn, K. C. "New Challenges in the Great Lakes States to Banning Phosphorus in Detergents." *Water Pollution Control Federation Journal,* 54 (1982):1,342–345.

Interagency Task Force on Acid Precipitation. *Annual Report, 1982, to the President and Congress*. Washington, DC: National Acid Precipitation Assessment Program, 1983.

Lantzy, R. J. and F. T. McKenzie. "Atmospheric Trace Metals; Global Cycles and Assessment of Man's Impact." *Geochemica et Cosmochemica Acta* 43 (1979):511–21.

Lipschutz, R. D. *Radioactive Waste: Politics, Technology, Risk.* Cambridge, MA.: Ballinger, 1980.

Matthes, G. *The Properties of Ground Water*. New York: John Wiley, 1982.

Olivieri, A. W. and D. M. Eisenberg. "A Methodology for Ranking the Relative Risk of Groundwater Contamination from Hazardous Material Sites," in *Environmental Engineering,* M. Pirbazari and J. S. Devinny, editors. Proceedings of 1984 Specialty Conference, American Society of Civil Engineers, 1984, pp. 187–92.

Pye, Veronica I. and Jocelyn Kelley. "The Extent of Groundwater Contamination in the United States," in *Groundwater Contamination*. Washington, DC: National Academy Press, 1984, pp. 23–33.

Pye, Veronica I., Ruth Patrick, and John Quarles. *Groundwater Contamination in the United States*. Philadelphia: University of Pennsylvania Press, 1983.

Rickert, D. A., W. G. Hines, and S. W. McKenzie. "Methodology for River-Quality

Assessment with Application to the Willamette River Basin, Oregon." *U.S. Geological Survey Circular 715–M,* 1976.

Schindler, D. W. "Evolution of Phosphorus Limitations in Lakes." *Science* 195 (1977):260–62.

Sevier, Robert and Werner Stumm. "Quality of Water—Surface and Subsurface," in *Scientific Basis of Water-Resource Management.* Washington, DC: National Academy Press, 1982, pp. 58–71.

Smith, R. A. and R. B. Alexander. "A Statistical Summary of Data from the U.S. Geological Survey's National Water Quality Networks." *U.S. Geological Survey Open-File Report 83–533,* 1983.

Trussell, A. R., M. F. Umphres, R. R. Trussell, and M. C. Kavanaugh. "Synthetic Organic Contaminants: Problems and Solutions," in *Environmental Engineering,* M. Pirbazari and J. S. Devinny, editors. Proceedings of the 1984 Specialty Conference, American Society of Civil Engineers, 1984, pp. 439–41.

U.S. Bureau of Reclamation. *Colorado River Improvement Program, Status Report, January 1983.* Denver, CO: U.S. Bureau of Reclamation Colorado River Water Quality Office, 1983.

U.S. Congressional Research Service. *Resource Losses from Surface Water, Ground Water and Atmospheric Contamination: A Catalog.* Washington, D.C.: U.S. Senate Committee on Environment and Public Works, 1980.

U.S. Council on Environmental Quality. *Contamination of Ground Water by Toxic Organic Chemicals.* Washington, D.C., 1981.

U.S. Environmental Protection Agency. *Damages and Threats Caused by Hazardous Material Sites.* Washington, D.C.: Oil and Special Materials Control Division Report, 1980.

U.S. Environmental Protection Agency. "Maximum Contaminant Levels (subpart B of part 141, National Interim Primary Drinking Water Regulations)." *U.S. Code of Federal Regulations,* Title 40, Parts 100 to 149, revised as of July 1, 1982, pp. 315–18.

U.S. Environmental Protection Agency. *A Ground-Water Protection Strategy for the Environmental Protection Agency, Draft.* Washington, D.C., 1984.

U.S. General Accounting Office. *Federal and State Efforts to Protect Ground Water.* Washington, D.C.: Report to the Chairman, Subcommittee on Commerce, Transportation and Tourism, Committee on Energy and Commerce, House of Representatives, GAO/RCED–84–80, 1984.

U.S. Geological Survey. "Water-Quality Issues," in *National Water Summary 1983—Hydrologic Events and Issues.* Washington, D.C.: U.S. Government Printing Office, Water Supply Paper 2250, 1984, pp. 45–63.

U.S. National Academy of Sciences Geophysics Study Committee. "Overview and Recommendations," in *Groundwater Contamination.* Washington, D.C.: National Academy Press, 1984, pp. 3–20.

U.S. Nuclear Regulatory Commission. *A Study of the Isolation System for Geologic Disposal of Radioactive Wastes.* Washington, D.C.: National Academy Press, 1983.

Wilson, John T. and J. F. McNabb. "Biological Transformation of Organic Pollutants in Groundwater," *EOS: Transactions, American Geophysical Union,* 64 (1983):505.

Wood, Eric F., Raymond A. Ferrara, William G. Gray, and George F. Pinder. *Groundwater Contamination from Hazardous Wastes.* Englewood Cliffs, NJ: Prentice-Hall, 1984.

3

IDENTIFYING KNOWLEDGE GAPS ON LAND/SOIL PROCESSES: HAZARDOUS SUBSTANCE AND THE LAND/SOIL RESOURCE

Louis J. Thibodeaux

INTRODUCTION

Surely everyone agrees that humans, as well as other organisms, depend upon the land/soil environment for survival. To adequately meet the needs for food, water, and shelter the land/soil resource must be used and manipulated in various ways. Clearly, effective management of this resource is fundamental to supplying society with all of its vital goods and services while maintaining a suitable environment.

Increasing demands on diminishing land/soil resources call for enhanced understanding of land's resilience and continuing productive potential under differing use scenarios. Land/soil uses have changed significantly in the past 50 years. New uses have been forthcoming in recent years that include the placement of toxic, hazardous, and long-lived waste onto and into the land/soil system.

For thousands of years land/soil has been used for agricultural and silvicultural ends. Over this period efforts at coaxing increased yields of food and fiber were limited to the use of natural or "organic" fertilizers and natural means of pest controls. The use of these so-called natural substances and processes does not imply that there were no abuses. The advent of modern chemistry and the use of these anthropogenic substances has, however, changed dramatically the nature of the impact.

The land/soil is being impacted by substances which can be classified as unnatural to the system. Since World War II there has been a systematic increase in the number and quantity of petroleum-based halogenated chemicals used to enhance food and fiber production. Organo-metallic compounds, inorganics, and metals have also been on the increase during this period. Applications have typically been in trace quantities. Hazardous wastes[1] have, by necessity, also been produced in trace quantities. The ''application'' of this material onto and into the land/soil system has typically been at high concentration levels.

The hazardous waste management problem and its impact upon the land/soil resource has recently begun to receive attention. National awareness of solid and liquid hazardous waste problems increased dramatically in the mid to late 1970s, with widespread concern over mismanaged waste, indiscriminate dumping of uncontainerized liquid waste, and infrequent but highly visible transportation accidents. It has become clear that even well-intentioned and accepted management of hazardous waste, particularly the use of landfills, surface impoundments, and lagoons, could result in a substantial threat. This threat results from the potential, but difficult to assess, slow leakage of waste constituents, or leachate (resulting from the interacting of water of other solvents and waste), through the soil and into the groundwater (Johns et al., 1983).

Within the capabilities of the author this chapter represents a freestanding assessment of the status of the scientific and technological knowledge base and gaps plus research and development recommendations about substantive areas on the specific topic of land/soil processes related to the use and misuse of hazardous substances in and on the land/soil system. A number of stakeholder groups have a direct interest in short- and long-term research on this topic.

Industry groups, representing the agricultural, silvicultural, chemical processes, and waste management, realize that long-term research on the topic is important to the continued maintenance of environmental quality and human health. This assessment is based in part upon the acceptance by industry of the spirit of regulatory reform that has occurred with respect to environmental concerns in recent decades. A typical response: ''While we might not agree with every facet of the current laws or technical aspects of regulations covering the management of hazardous waste, we believe that conceptually such regulations were certainly necessary and desirable'' (Lederman and Daniels, 1974). Regulations such as CERCLA have involved a tax on industry. A part of this collection has been routed to research and development aimed at land/soil processes and hazardous waste.

Environmental and public interest groups have also been active to point out the need for increased knowledge about land/soil processes and hazardous waste. Research on the utilization of the treatment and assimilative capacity of all phases of the global environment was considered justified by the committee. With adequate testing, monitoring, and regulation, the successful utilization of the soil's assimilative capacity could be a viable and justified technological alternative in the management of industrial hazardous waste (NRC, 1983).

STATUS OF SCIENCE AND TECHNOLOGY KNOWLEDGE BASE AND KNOWLEDGE GAPS

It is convenient to classify chemical concentrations in soils as either high or low. High concentration levels are taken to be those in which the chemical contaminant is present at the 5 percent level or greater. Establishing such a definite level is sure to cause controversy because an ''effect'' basis must be adopted if the choice is not to be arbitrary. There are many effect levels one could choose. This choice of 5 percent is based on physical effects only. With this level of contaminant present the natural soil phase begins to be influenced significantly by the presence of this foreign substance. For example, the volume basis for concentration needs a proper definition, and transport coefficients are not constant (Thibodeaux, Wolf, and Davis, 1983). Individuals who consider soil as ''living matter'' would want to define levels with respect to biodepressive effects.

LOW TO HIGH CONCENTRATIONS OF HAZARDOUS SUBSTANCES IN LAND/SOIL PROCESSES

The following sections will contain a review of land/soil processes with respect to hazardous substances that range in concentration from very low to very high.

Pesticide Applications to Land/Soil

Pesticide applications to soil surface for purpose of pest control are typically in the low concentration range. In some respect the application of pesticides on soils is special in that this is an application of hazardous chemical to soil for a net beneficial effect. The benefit is the control of a pest in order to increase food or fiber production. In contrast, the application of hazardous wastes to soils is often done to ''treat'' the hazardous substance and destroy or change the chemical itself.

Effects of the application of pesticides on soils has received much attention. These types of applications have long been identified as potentially hazardous to segments of the soil environment and other parts of the ecosystem and have received attention in the scientific literature. One particularly important publication is *Residue Reviews—Residues of Pesticides and Other Contaminants in the Total Environment*. A recent edition addresses the question of the types and magnitudes of chemical hazards on the biotic compartments of the soils and compares these to naturally occurring hazards. The chemicals of interest were agrochemicals (Domsch, Jagnow, and Anderson, 1983). The authors point out that a wealth of data is available but it contributes only partially to our understanding of, and capacity to assess, hazards to soil. They argue that it should be possible to develop ecologically based "yardsticks" to provide government regulators with a means of identifying the use of "critical" environmental chemicals.

This work on the effect chemicals have on soil microorganisms typifies our base of knowledge on the low level contaminants. Obviously we know much about the effects of pesticides on soil, but there is much we do not know. Green (1981) observes that considerable progress has been made in modeling pesticide transport processes for well-defined systems such as laboratory columns, but accurate field-scale predictions are generally still beyond our grasp.

Another specific example that indicates to me the level of our scientific base is the incapability of predicting vapor pressure exerted when various levels of pesticide are applied to a soil. A recent review article attempts to put this subject in a state of organization (Spencer et al., 1982). At very low levels of soil loadings in the range of 50 ppm, dieldrin, for example, exerts its pure component vapor pressure. As the water content is reduced, higher levels are needed to yield the pure component vapor pressure. The critical loading rates at which chemicals exert their pure component vapor pressure apparently vary with chemical species, soil type, temperature, and so on, in a yet-to-be determined fashion, making predictions extremely difficult. The availability of generalized chemical vapor pressure/solid loading models would have a significant impact on the current efforts to assess the vapor mobility of 2, 3, 7, 8–TCDD which exist at low levels (by above definition) on surface soils throughout the state of Missouri (Freeman and Schroy, 1984).

Chemical transport mechanisms in the air boundary layers immediately above the soil surface are still undergoing reevaluation and study (Glotfelty et al., 1983). It appears from recent field studies involving a mixture of pesticides, that the individual transport coefficients of the species are unrelated to the conventional molecular weight and structure properties of the molecules. The air side of the land/soil interface is exceedingly simple, in many respects, compared to the soil side. The knowledge gaps are exceedingly greater in number on the soil side.

The knowledge gaps are obviously critical in some area for low level chemical contaminants in land/soil systems. The gaps are in the areas of chemical equilibrium, transport, and reaction in the upper layers of the soil. Predicting the behavior and fate of pesticides and other low level contaminants, in a general qualitative sense and in specific quantitative calculations, is hampered by a general lack of critical information. Specific details of the exact nature of some of these gaps will be covered later in this manuscript.

Hazardous Waste Application to Land/Soil

The concentration levels of hazardous waste present on soils are usually higher than the levels associated with pesticide applications. This is due mainly to costs.

Table 3–1. Hazardous Waste Management Facilities During 1981[a]

Technology Type	Estimated No. of Sites	Estimated Total
		Waste throughput
Injection wells	114	3.5 E 9 gal
Landfills	270	8.3 E 6 tons
Land treatment	148	8,600 acres
Surface impoundments	1,096	28.8 E 6 sq. yds.
Waste piles	312	13.2 E 6 cu. yds.
Incinerators	317	227 E 6 gal
Storage containers	5,652	57 E 6 gal
		Estimated capacity
Storage tanks	2,280	303 E 6 gal
Treatment tanks	1,951	3.1 E 9 gal
Total	7,785	

[a]adapted from Johns et al. (1983).

In the case of pesticides, the application rate is cost effective if the minimum required for effective pest control is used. In the case of hazardous waste treatment, the application rate is cost effective if the maximum assimilative capacity of the soil is used. Hazardous waste management operations that involve land/soil processes as an element treatment include landfills, land treatment, surface impoundments, and waste piles. A compilation of hazardous waste management facilities operating in 1981 was performed by Johns et al. (1983). Table 3–1 shows an estimated total of 7,785 hazardous waste management facilities in nine technology classes. Of these 1,826 involve the land/soil system directly in the "treatment" scheme.

The above estimate does not reflect CERCLA (Comprehensive, Environmental, Response, Compensation and Liability Act of 1980) sites. These sites are those that contain hazardous waste that may require cleanup. National estimates of the number of sites have been provided by two studies, by an EPA consultant and the Chemical Manufacturers' Association. The respective estimates range from 50,000 sites to 4,800 sites (Johns et al., 1983). A more recent estimate places the number between 17,000 to over 22,000 (Ruckelshaus, 1984). The hazardous waste associated with these uncontrolled sites involved extensive land/soil contamination. The disposal of the cleaned up material involves extensive use of landfills, landfarms, and surface impoundments.

A typical CERCLA activity is the recent USEPA-approved cleanup of Waukegan Harbor, Illinois (Bernard, 1984). PCB-contaminated waste oils were allegedly discharged into the harbor from 1956 to 1976. Concentrations were found over 10,000 parts per million (1 percent) in bottom sediment. A coffer dam and dredging will prevent the spread of the PCB-contaminated sediment. Dredged, contaminated sediment with concentration of 700,000 ppm or greater will be sent to a treatment plant for fixation. Fixed solids will be disposed of at a chemical waste landfill. Dredged sediment with lesser concentrations will be treated in various ways including dewatering in clay-lined lagoons followed by capping.

TREATMENT AND DISPOSAL FACILITIES

The following sections contain brief descriptions of common treatment and disposal facilities that involve extensive land/soil processes.

Surface Impoundments

This facility is usually aqueous and resides upon the land. In general a surface impoundment is an open vessel, usually of earthen material construction, on the surface of the ground, into which is placed liquids and sludges. Surface impoundments are used for operations such as storage, settling of solids, treatment, and evaporation. Because of the intimate connectivity with the land/soil system it is included in this review.

Biological treatment in surface impoundments involves the degradation of organics either by an indigenous microbial population or by organisms adapted to act specifically on a compound or group of compounds. Both aerobic and anaerobic processes are used. The most common biological treatment methods include activated sludge, aerated lagoons, trickling filters, stabilization ponds, and anaerobic digestion. These processes are generally used for the treatment of liquids or slurries.

A recent report highlights some of the knowledge gaps with respect to the biological treatment of hazardous industrial waste (NRC, 1983). To realize the full potential of biological treatment for the control and detoxification of hazardous industrial waste, additional information is needed on the mechanisms of hazardous waste removal in biological systems, on unconventional biological treatment processes that can be used, and on approaches to defining the lower levels of toxicity of hazardous industrial waste. Further specific details of research needs are contained in that report.

Many biological treatment operations are performed in shallow earthen basins. The biological processes are highly related to soil microbial processes, and therefore similar knowledge gaps are likely to exist. The major research needs highlighted in the NRC report were:

- Identification of the removal mechanisms for components of hazardous industrial wastes so that better treatment processes can be developed and so that transfer to toxic pollutants from one medium to another will be minimized.

- Evaluation of aerobic thermophilic processes for detoxification of hazardous industrial wastes; such processes may be technically and economically feasible with concentrated organic industrial wastes.
- Evaluation of the potential of genetically-adapted microorganisms to detoxify specific hazardous industrial wastes.
- Determination of analytical methods, short of fish bioassay methods, that can provide measures of detoxification when biological treatment processes are used; such methods should be capable of routine use.

Landfarms

Land treatment is an alternative name. It does not involve the production of crops but the use of soil system of a vegetation-soil system as the ultimate receiver of a waste. Figure 3–1 shows such field operations. The hazardous waste material applied to the land may be slurries, sludges, untreated waste, residues, or solid waste. Typically, the wastes are mixed with or applied to the surface (0–1 feet) of land. Chemical and biological reactions then break down a portion of the waste, adsorption and fixation occur for other portions, and controlled migration is allowed for certain anionic inorganic fractions.

Because of its cost-effectiveness and the idea that the land area used is not rendered permanently restricted to further use, land treatment may result in some abuses and excesses and therefore problems. A case in point is the use of genetic engineering of waste-specific microbes for the in situ degradation treatment of hazardous components. Reports of treatment removal of selected hazardous components by such operations are possibly overrated because of the lack of consideration of competitive removal mechanisms. Volatilization of original materials and bio-reaction products is a rapid and effective removal mechanism which is not treatment in the ultimate disposal sense (Thibodeaux and Hwang, 1982). This aspect has not been considered in the NRC list of major research and development needs (NRC, 1983). The listed needs of that report are:

- Substantial expansion of the system design data base for land treatment with an emphasis on transferable laboratory-scale information and verification in selected field-scale systems.
- Documentation of the economics of full-scale hazardous waste land treatment systems, including costs of the individual field components of the total system.
- Technical and economic consideration of process modifications or pretreatment in combination with land treatment to yield a lower total system cost than treatment alone.

Additional knowledge gaps and research needs related to chemicals on or near the soil surface are presented by Thibodeaux, Wolf, and Davis (1983), and these are reviewed briefly in the following paragraphs. These aspects involve solubility, vapor pressure, and sorption.

Water plays a dominant role in the soil environment. It appears that if sufficient soil water is present and dilute solutions exist, then Henry's constant can be used to obtain partial pressures for solution concentrations up to the solubility limit. Spencer, Farmer, and Jury (1982) in a recent review observed that the vapor pressures of lindane, DDT, and trifluralin dropped to very low values when the water content was decreased below that equivalent to approximately 1 molecular layer, presumably by adsorption due to an increased competitive advantage. Significant differences in vapor density (or pressure) occur for dieldrin when soil water is reduced from 3.94 percent to 2.1 percent.

This occurrence and the observations that the vapor emission rate from soils increases dramatically under certain conditions of soil moisture suggest that our knowledge of the equilibrium physical chemistry processes of dilute chemical soil mixtures in the range of unsaturated to ''bone-dry'' soilwater conditions is lacking. Cupitt (1980) used the Brunauer, Emmett, and Teller (BET) modified Langmuir adsorption theory to estimate the vapor pressures of toxic chemicals adsorbed onto ''bone-dry'' aerosols. He gives no data to support the validity of the BET model. Bailey and White (1970) suggest that the same model can be used for pesticides on soils.

Various sludges generated by industrial operations and wastewater treatment operations contain waste materials that are potentially hazardous. Three general types of organic sludges or organic matter can be identified: (1) natural organic matter produced from normal soil processes such as biological decay products form biomass material consisting of grass, leaves, agricultural residues, and other organic material; (2) biosludges produced from microbial cultures of wastewater treatment plants such as activated sludge, anaerobic digesters, primary filtration, and others; and (3) oily/chemical sludges from petroleum, petrochemical, or organic chemical manufacturing operations which include sources such as American Petroleum Institute (API) oil separators, tank bottoms, and still bottoms.

Partial pressures of selected chemicals from some of the chemical process sludges may be estimated by conventional techniques used in process design as outlined by Perry and Chilton (1973). A simplified version of this design was employed by Thibodeaux and Hwang (1982) in modeling the air emission from a petroleum landfarming operation. The vapor pressure of the volatile species ''dissolved'' in the oily sludge can be estimated with Raoult's law. Raoult's law applies well if solute and solvent have no heat on mixing and no volume change on mixing. These ''ideal solution'' rules are likely valid for cases such as bensene in API sludge or in still bottom sludges; however, these specific equilibrium systems have little support data. For vapor pressures of volatiles above biosludges and natural organic matter ''solvents,'' no studies have been found.

The partition coefficient, K_p, for trace chemicals between earthen solid, either soil or sediment, is defined as the ratio at equilibrium of the concentration on the solid to the concentration in water. In soils and sediments the adsorption of aromatic hydrocarbons and chlorinated hydrocarbons as expressed by K_p is directly related to the organic carbon content of the adsorbent. In such systems, it is convenient to use the term, K_{oc}, which is simply $_p$ divided by the organic carbon content of the adsorbent. The K_{oc} has also been shown to be closely correlated to the octanol/water distribution coefficient, or K_{ow}, for several compounds (Karickhoff et al., 1979; Chiou, Porter, and Schmedding, 1983). A similar conclusion was reached by Brown and Flagg (1981) in their study of nine chloro-*s*-triazine and dinitroaniline compounds.

The partition coefficient will likely need to be generalized to include high chemical concentrations. It is unlikely that a simple ratio will suffice. The present generation of useful correlations covers a very narrow range of conditions that include organic chemicals in surface soils which have considerable organic matter content. These correlations are typified by the recent work of Chiou, Porter, and Schmedding (1983).

The spectrum of research scenarios for the adsorption (or solution) of chemicals onto (or into) soil systems needs to be broadened. The work of Anderson, Brown, and Green (1982) on the influence of adsorbed organic fluids on the change in permeability of clay soils highlights this need.

This zone is characterized by high chemical concentrations and possibly additional liquid phases that overwhelm the subsoil adsorption capacity for the leaching constituents. For a given subsoil system the information currently available in the literature does not allow one to estimate the quantity of chemical adsorbed without performing simulation experiments. Adsorption is largely based upon the organic matter content of the soil, which may not exist for subsoil systems, or if it does, the high chemical concentration overwhelms the "solution" or solvent capacity of the organic matter.

Recent research work involving the sorption of pesticides in the presence of co-solvents, such as water-methanol and water acetone, in subsurface environments appears to be a realistic approach in extending the retardation factor concept to include such high concentration mixtures (Rao et al., 1983).

It is also possible that certain classes of organic chemicals such as phenolic compounds may undergo polymerization reactions as described by Wang et al. (1978) and result in the formation of humiclike materials which have properties similar to soil organic matter (Martin, 1972). Should this occur, it would be possible to develop a synthetic soil horizon with increased "organic matter" levels which would more nearly approach the description of an A rather than a B horizon. This could result in increased adsorption of the chemical compound.

Spill Sites, Contaminated Land, Processing Plant Sites

Spill sites and contaminated land introduce scenarios of high chemical concentrations related to sorption onto surface soils. Freeze and Cherry (1979) present the stages of migration of oil seeping from a surface source and define a residual oil saturation. This parameter is defined as a stable stage when an oil spill on a soil surface is held in a relatively immobile state in the pore spaces. Experiments in our laboratory (Altenbaumer et al., 1982) with the organic liquids propanol, acetone, ethylene glycol, crude oil, and motor oil suggested that surface tension was the most important independent variable affecting the residual saturation. Residual saturation is the volume of liquid immobilized divided by the initial soil pore volume. Values ranged from 0.33 to 0.75, indicating that 33 to 75 percent of the available soil void volume was occupied by the liquids. The residual saturation for water was 0.26 for this soil which was 1.42 percent organic matter.

The above range of soil contaminant conditions, trace contaminants to concentrated leached to pure fluids, is the state of many important hazardous substance problems for which adsorption equilibrium information is almost totally lacking. A comprehensive approach including theory and experiments along the lines of Dexter and Pavlou (1978) seems to be an appropriate first step. Incorporated into this approach are functional groups and chemical structure parameters which must be included in any comprehensive approach due to the varied nature of the organic chemicals involved.

Landfills

A dictionary definition of a landfill is "a device for disposal of garbage or rubbish by burying in the ground." There are several classes of landfills, and three will be described here that relate to the disposal of hazardous chemicals.

Sanitary Landfills

These landfills are constructed to receive primarily domestic solid waste. Other waste such as treatment plant sludges, building materials, grass and tree cuttings, ashes, and so on, are also acceptable in sanitary landfills.

The fundamental unit or building block of a sanitary landfill is a "cell," which contains solid waste and is completely closed on all sides by soil. The depth of a cell can vary between 60 cm to 4.6 m. It is constructed by compacting waste on a slop in several layers. Soil is placed over all exposed waste material at the end of each working day. The thickness of this daily soil cover is generally never less than 15 centimeters when compacted. The above description is that of Pohland and Engelbrecht (1976).

Unsecure Chemical Landfill

Landfilling has long been the most common method for disposal of hazardous waste. Burial as in a conventional sanitary landfill, which differs from a secured chemical landfill primarily in the degree of protection against leaching, involves simple technology. Typically, a hole is dug, unconsolidated sludge and drums of chemicals are placed in it, and the hole is filled and covered with clay to keep out rain and other water.

Secure Chemical Landfills

These are a new generation of landfills of which the following is an example of a good type of operation. The landfill cells, 150 meters long and 150 meters wide, are excavated in dense clay and fitted with liners of impervious, reinforced synthetics, which are then covered with a layer of about 1 meter of clay. The main landfill cell is divided by clay barriers into a number of subcells for specific types of hazardous waste. Drums of PCBs and other chlorinated hydrocarbons might be kept in one cell, for example, precipitated metal hydroxides in another, and organic sludges in a third. When it is filled with waste, in about 12 to 18 months, the pit is capped with another synthetic liner and a layer of clay. This description of secured chemical landfills is extracted from Maugh (1979). Figure 3–1 illustrates a landfill and chemical vapor transport pathways.

The present status of the technology of hazardous waste landfills for "permanent storage" has been reviewed recently (NRC, 1983). The report emphasizes secure landfills because they are currently the major method of disposal of hazardous waste. The objective of secure landfills is the physical (hydraulic) isolation of waste from the environment. In simple terms, a secure hazardous waste landfill is designed to provide long-term entombment of waste, and one of its prime purposes is to prevent contamination of groundwater by preventing "leaching" of material from the site. The report states that secure landfills must be used until more advanced methodologies become viable industrial processes; however, for the longer term, other hazardous waste disposal options must be pursued and developed to minimize burial of hazardous material. It therefore seems appropriate to address the land/soil processes of this technology since it will likely be with us for the rest of this century.

The NCR report listed and discussed specific research needs. In brief summary this list includes:

- Further develop landfill design and operation techniques. This includes research on cover design and liner performance (natural, synthetic). Emphasis should be on leachate.
- Further develop closure-leachate collection, analysis, and treatment including sampling (wells) and leachate treatment.
- Study site monitoring including pollutant movement and sampling and interactions of soil, waste, and so on. Develop sampling and preservation techniques for groundwater/leachates.
- Develop waste pretreatment or chemical stabilization, including fixation, encapsulation, and solar evaporation. Specific research needs include identification of new encapsulation and solidification materials, determination

Figure 3–1. Chemical Vapor and Bio-gas Movement In and Around a Landfill

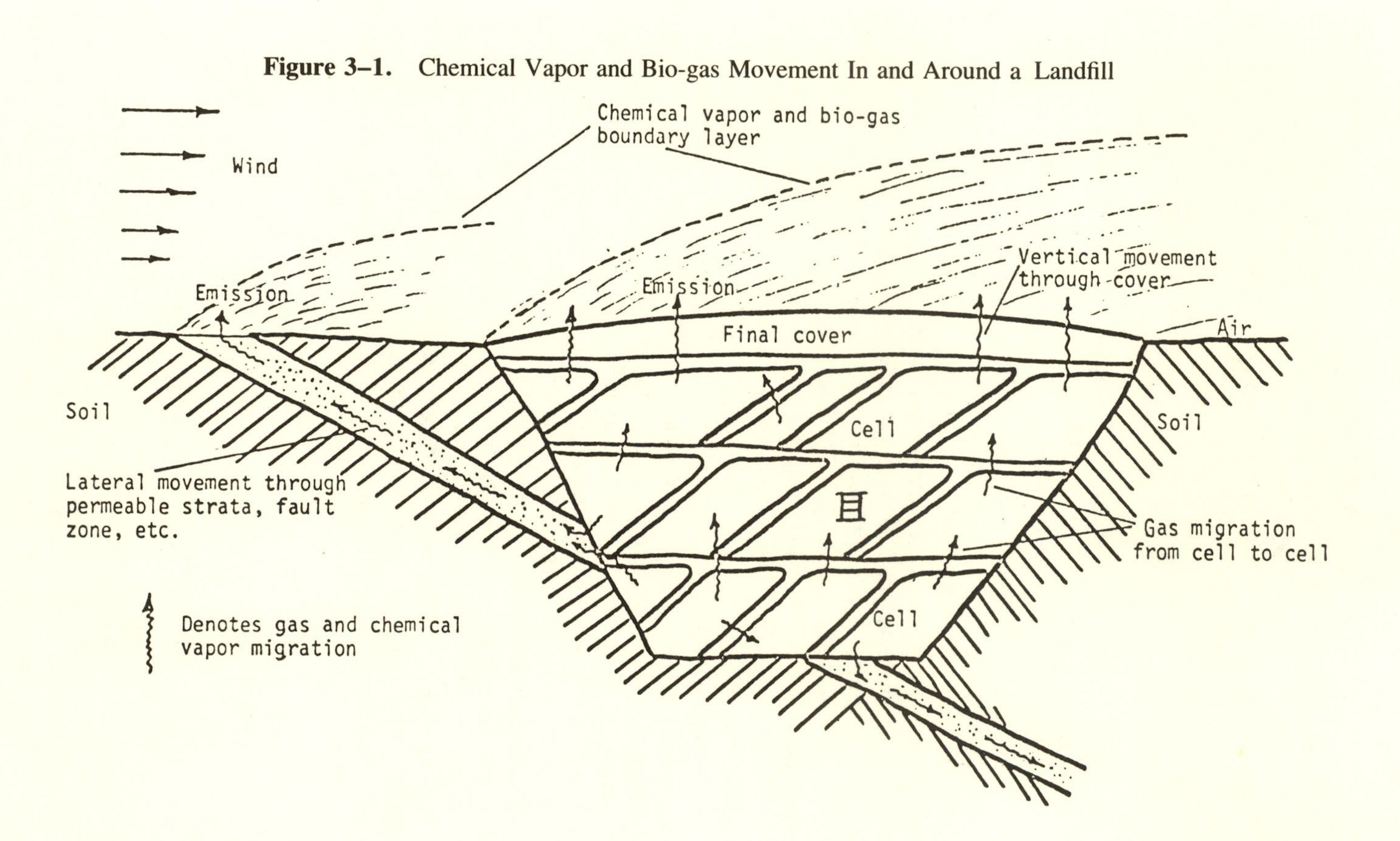

of long-range stability with respect to leaching, and determination of ability to withstand "weathering" after prolonged periods of time.

- Perform model development that will project long-term (500 y) care.
- Develop standard waste (solid and semisolid) sampling procedure that include collection, preservation, and storage.
- Perform pollutant migration studies in order to: use soil as a predicable attenuation medium, understand the processes of migration from sites, verify with laboratory and field studies, and develop mathematical hydrogeological simulation models for predicting the movement of leachate (solutes) in both saturated and unsaturated soils.

Considering the number of research items and the emphasis of the investigations, it seems very premature, from a technical point of view, to strongly recommend secure landfills as a means of long-term entombment. The above list is incomplete in two very important aspects: (1) pollutant migration is limited to leachate processes, and (2) no consideration is given to understanding the properties and behavior of the waste inside the fill cells. It continues to amaze me that otherwise competent engineers/scientists continue to address waste and waste processes from a technical perspective that would be absolutely unsatisfactory in the highly competitive world of product research and development.

Investigations of transport processes for pollutants from landfills has been mostly concerned with leachate and groundwater. Research in our laboratory has considered the volatile emission problem (Thibodeaux, Springer and Hildebrand, 1984). It appears that there are other processes, including capillary effects, that are quite effective in causing chemical transport both within the cells and through the liners (Cairney, 1982). Experiments established that, in drought conditions, a clay layer placed over a waste material can promote significant upward migration of pollutants.

The migration of chemical vapors in the unsaturated zone near landfills is not understood competely. It has been observed that there is apparent migration of volatile organic vapors and re-dissolution in groundwater at a point removed from the landfill. The importance of this transport process around landfills need be investigated through modeling and experimental work. We know very little of the behavior of organic chemicals within the cells of landfills. Liquid and sludge wastes are mixed with fill material to create a "solidified" mass that can be managed with conventional earth moving equipment. This is a common operation employed in cleanup of present day "Superfund" sites. There appears to be a lack of information on the binding of organics in this "solidified" state (Tittlebaum et al., 1983), equilibrium (air and water) sorption properties, and reactivity, both chemical and biological. The NRC report mentions the "small" generators and their practice of disposing of hazardous industrial waste in sanitary landfills, and that if enough waste is aggregated in on such landfills then leachate, groundwater problems could result. The climate within the cell environment of sanitary landfills is much more active and will likely cause enhanced transport and reaction effects when compared to the secure landfill cells. The enhancement effects are due largely to the presence of sufficient moisture and cellulose refuse materials.

Research needs involving waste scenarios of high chemical concentration have been listed by Thibodeaux, Wolf, and Davis (1983) and include:

- Laboratory simulation of leachate migration processes for high chemical concentration in saturated soils;
- Model reformation based upon the observed mechanisms to account for the density stratification and the presence of two phases in saturated soil;
- Determine the effective diffusion coefficient for the gas phase and the liquid phase in unsaturated porous media;
- Study the convection processes in the liquid phase as driven both by evaporation of water at the soil surface and the capillary forces from liquid waste below the surface;
- Study the convection processes in the gas phase as driven by internal gas, atmospheric pressure pumping, and water evaporation beneath the surface;
- Thermal energy flux rates within the top 30 to 50 cm of the surface;
- Equilibrium sorption on "air-dry" soils, including polar and nonpolar species in single, binary, and tertiary mixtures;
- Equilibrium sorption on moist soils concerning the behavior of isotherms for water contents between "air-dry" and 10 percent water; and
- The physicochemical nature of organic refuse material at the waste disposal sites.

Additional research recommendations that would increase our knowledge base concerning land/soil processes associated with hazardous waste in landfills include:

- Perform archaeological investigations at early co-disposal sites to learn from the residue things as extent of movement, reaction, sorption, fixation, and so on;
- Develop test protocols for simulation long-term reactivity and products;
- Study in-cell process both in the laboratory and field under present and proposed secure landfill operating conditions.

SUMMARY

This study is mainly concerned with land/soil processes and hazardous waste. The application of large amounts of waste to yield high concentration conditions on land/soil gives rise to a new set of problems and questions. Examples and scenarios have been presented that are related to hazardous waste in surface impoundments, spill sites, dump sites, landfarms, and landfill.

Knowledge gaps were exposed. This process was aided, in part by the recent publication of two key documents that address various aspects of hazardous waste treatment/disposal and hazardous waste behavior in the natural environment. Research recommendations from those documents were reviewed and supplemented with recommendations resulting from the knowledge gap analysis exercise demanded by this study. The cumulative lists, made of specific items, suggest that much research needs to be done to raise the level of understanding and quantification of land/soil processes associated with hazardous waste.

NOTE

1. The term "hazardous waste" means a solid, sludge, or liquid waste substance, or combination of waste substances, which because of its quantity, concentration, or physical, chemical, or infectious characteristics, may (a) cause, or significantly contribute to, an increase in mortality, or an increase in serious irreversible, or incapacitating reversible, illness, or (b) pose a substantial present or potential hazard to human health and the environment when improperly treated, stored, transported, or disposed of or otherwise managed (Johns, 1983).

REFERENCES

Altenbaumer, T., R. Kew, B. Matthews, and L. Wright. "A Study of Contaminated Soil Volumes Resulting from Organic Liquid Spills." Senior Project Report, Dept. of Chem. Eng., Univ. of Arkansas, Fayetteville, 1982.

Anderson, D., K. Brown, and J. Green. "Effect of organic fluids on the permeability of clay liners." EPA 600/982 002. *Proc. 8th Ann. Res. Symp., Land Disposal of Hazardous Waste,* Ft. Mitchell, KY, 1982, p. 179.

Bailey, G. W. and J. L. White. "Factors influencing the adsorption, desorption, and movement of pesticides in soil." *Residue Rev.* 32 (1970):29–92.

Bernard, H. (ed.). *Focus,* Silver Springs, MD: Hazardous Material Control Research Institute, July 1984, p. 6.

Brown, D. S. and E. W. Flagg. "Empirical prediction of organic pollutant sorption in natural sediments." *J. Environ. Qual.* 10 (1981):382–86.

Cairney, T. "In situ reclamation of contaminated land: The problem of safe design." *Public Health Engineer* 10, 4 (1982):215–18.

Chiou, C. T., P. E. Porter, and D. W. Schmedding. "Partition equilibria of nonionic organic compounds between soil organic matter and water." *Environ. Sci. Technol.* 17, 4 (1983):227.

Cupitt, L. T. "Fate of toxic and hazardous materials in the air environment." EPA–600/3–80–084. Research Triangle Park, NC: U.S. Environmental Sciences Res. Lab, 1980, pp. 12–17.

Dexter, R. N. and S. P. Pavlou. "Distribution of stable organic molecules in the marine environment: Physical chemical aspects chlorinated hydrocarbons." *Marine Chem.* 7 (1978):67–84.

Domsch, K. H., G. Jagnow, T. H. Anderson, "An ecological concept for the assessment of side-effects of agrochemicals on soil microorganisms." *Residue Reviews* 86 (1983):65–105.

Freeman, R. A. and J. M. Schroy. "Environmental Mobility of Dioxins." *Proc. 8th ASTM Aquatic Toxicology Symp.,* Ft. Mitchell, KY, April 15–17, 1984.

Freeze, R. A. and J. A. Cherry. *Groundwater*. Englewood Cliffs, NJ: Prentice-Hall, 1979, pp. 444–47.

Glotfelty, D. E., A. W. Taylor, W. H. Zooler. " Atmospheric depositions of vapors: Are molecular properties unimportant?" *Science* 219 (1983):843.

Green, R. S. "Forecasting pesticides mobility in soils: Dispersion and adsorption considerations." USA-USSR Symposium Predicting Pesticide Behavior in the Environment, Yerevan, USSR, Oct. 1981. USEPA Environmental Res. Lab., Athens, GA. (EPA–600/9–84–026).

Johns, L. S., et al. *Technologies and Management Strategies for Hazardous Waste Control*. Washington, D.C.: Office of Technology Assessment, March 1983.

Karickhoff, S. W., D. S. Brown, and T. A. Scott. "Sorption of hydrophobic pollutants on natural sediments." *Water Res.* 13 (1979):241–48.

Lederman, P. B. and S. Daniels. *Commentary on Hazardous Waste Management*. New York: Govt. Programs Steering Comm., Am. Inst. Chemical Engineers, May 1974.

Martin, J. P. "Side effects of organic chemicals on soil properties and plant growth." In C.A.I. Goring and J. W. Hamaker (eds.), *Organic Chemicals in the Soil Environment*. New York: Marcel Dekker, 1972, pp. 733–92.

Maugh, T. H. *Science*, 204 (June 22, 1979):1295–298.

Pohland, F. G. and R. S. Engelbrecht. *Impact of Sanitary Landfills: An Overview on Environmental Factors and Control Alternatives*. Report for American Paper Institute, Feb. 1976.

National Research Council. *Management of Hazardous Industrial Waste: Research and Development Needs*. Pub. NMAB–398. Washington, D.C.: National Academy Press, 1983.

Perry, R. H. and C. H. Chilton. *Chemical Engineers' Handbook* (5th ed.). New York: McGraw-Hill, 1973.

Rao, R.S.C., A. G. Hornsby, and P. Nkedi-Kizza. "Influence of solvent mixtures on sorption and transport of toxic organic compounds in the subsurface environment." Paper 67, Env. Chemistry, Amer. Chem. Soc. Mtg., Washington, D.C., Aug. 28–Sept. 2, 1983.

Ruckelshaus, W. B. Statement before House Subcommittee on Commerce, Transportation and Tourism of Comm. Energy & Commerce, 98th Cong., 2d sess., H.R.4813 and 4915, Feb.–March 1984 (pp. 721–76).

Spencer, W. F., W. J. Farmer, and W. A. Jury. "Review: Behavior of organic chemicals at soil, air, water interfaces as related to predicting the transport and volatilization of organic pollutants." *Env.Tox.Chem.* 1 (1982):17–26.

Thibodeaux, L. J. and S. T. Hwang. "Landfarming of petroleum waste: Modeling the air emission problem." *Env. Progress* 1, 1 (Feb. 1982):42–46.

Thibodeaux, L. J., C. Springer, and G. Hildebrand. "Air Emissions of Volatile Organic Chemicals from Landfills: A Pilot-Scale Study." *Proc. 10th An. Res. Symp., U.S. EPA, Land Disposal of Hazardous Waste*, Ft. Mitchell, KY, April 3–5, 1984, pp. 172–80.

Thibodeaux, L. J., D. C. Wolf, and M. Davis. *Impact of High Chemical Contaminant Concentrations on Terrestrial and Aquatic Ecosystems: A State-of-the-Art Review*. Athens, GA: Athens-ERL, U.S. Env. Protection Agency, Sept. 1983.

Tittlebaum, M., R. Seal, and F. Cartledge. *Identification of Bonding and Interfering Mechanisms Associated with Stabilized/Solidification Hazardous Organic Waste*.

Final Report. Baton Rouge, LA: La. St. Univ. Hazardous Waste Research Center, 1983.

Wang, T. S. C., S. W. Li, and Y. L. Ferng. ''Catalytic polymerization of phenolic compounds by clay minerals.'' *Soil Sci*. 126 (1978):15–21.

4

ADVANCING KNOWLEDGE ON PROTECTION OF THE LAND/SOIL RESOURCE: ASSIMILATIVE CAPACITY FOR POLLUTANTS

Dr. Raymond C. Loehr

INTRODUCTION

Background

Soil is a natural resource that is used by man for food and fiber production and structural purposes. It also is used to assimilate the residues of human activity, such as human, animal, municipal, and industrial wastes.

There are three major "sinks" into which wastes ultimately are discharged: water, air, and land. Each has a specific assimilative capacity for the constituents that are discharged. Inadequate understanding of the water and air assimilative capacity contributed to the water and air pollution that has been obvious in the past decades. As the nature of the assimilative capacity of air and water has been established and the environmental impacts better understood, water and air quality criteria and standards have been developed to protect these resources and to assure that the assimilative capacity of these entities is not exceeded.

Current knowledge about the soil assimilative capacity is increasing but is less extensive than that for air and water. Reliable and quantitative predictions of contaminant movement in soil requires an understanding of the processes that control the transport, hydrodynamic dispersion, and chemical, physical, and biological changes of the contaminants in the soil. Increasing such understanding requires long-term research and is vital to the maintenance of environmental quality, human health, and the land and soil resource.

The assimilative capacity of the soil is poorly understood and that it has been exceeded is increasingly obvious [1–3]. Industrial waste lagoons, land disposal sites (landfills, pits, ponds, and lagoons), septic tanks, and agricultural practices can be sources of groundwater contamination. Incidents of groundwater contamination have occurred in every state.

The nature and characteristics of the wastes, the soil characteristics, the topography of a site, the climatic conditions, the vegetation grown at the site, and the use of a site affect the fate of contaminants added to a soil. If the constituents are mobile, there is a risk that they will affect the quality of surface and ground waters. The degree of risk caused by the contaminants will depend upon the concentration and toxicity of the contaminant as well as the degree of human and environmental exposure that will occur. Standards and criteria to assess the degree of risk associated with the land application of wastes and to protect the public and the environment are not extensive. Except for drinking water criteria and standards, most of the available environmental criteria and standards have been developed to protect air and water, and not soil, from adverse impacts.

Fortunately, guidelines that will protect the soil resource and avoid exceeding the soil assimilative capacity are being developed at the federal, regional, and state levels. Additional research to more clearly understand the transformations, transport, and fate of constituents added to the soil will accelerate the development and application of sound criteria, guidelines, and standards.

Objectives

The objectives of this chapter are to: (a) identify the factors that influence and determine the assimilative capacity of soils, (b) indicate the status of the knowledge, (c) identify research and development needs that can fill gaps in the knowledge base, and (d) provide recommendations concerning the priority of such research and development. The factors are relevant to land-based waste management processes such as landfills, land treatment, surface impoundments, and waste piles.

Emphasis in this chapter is placed on the parameters that are of major environmental concern (a) nutrients, primarily nitrogen, (b) pathogens, (c) metals, and (d) synthetic organics. Other factors such as salts and odors can be of concern when the soil is used to assimilate wastes; however, these constituents are not discussed in any detail. Emphasis also is placed on the transformations that occur when wastes are applied to the soil and on the knowledge base associated with transport of the added constituents.

The waste application rates and the design, operation, and monitoring of the land used for waste treatment and disposal are equally important to avoid exceeding the assimilative capacity of the site. Due to space limitations, these latter factors will not be discussed in detail. However, research and development needs related to these factors will be indicated.

MAGNITUDE OF THE PROBLEM

As the other major sinks for residue disposal—water (streams, lakes, estuaries, and oceans) and air—are constrained, increasing attention has been given to the use of the soil for such disposal. If the assimilative capacity of the soil *is not* exceeded, there can be a societal benefit from the reuse of the water, nutrients, and organics in the wastes. If, however, the assimilative capacity *is* exceeded, water contamination and the loss of the soil as a productive resource can occur.

The soil represents not only an appropriate treatment and stabilization medium for many wastes but also an opportunity to manage wastes and recycle organics and inorganics with a minimum of adverse health and environmental effects. Application of manure, sewage, wastewaters, and certain industrial wastes to land has occurred for centuries.

The terrestrial environment can deactivate and stabilize many wastes as a result of the physical, chemical, and biological mechanisms in the soil. As long as the assimilative capacity of a specific site is not exceeded, the waste constituents are degraded, immobilized, and not transmitted through the atmosphere, water, or food chain. The concern of the public, industry, and regulatory agencies is to avoid exceeding the assimilative capacity of a site.

Large quantities of wastes are discharged to the soil from homes, cities, and industries. Table 4–1 estimates the quantity from some of these sources. For other sources, there are no sound estimates of the amount that reaches the soil. The basis for these estimates and information on other sources are noted in subsequent paragraphs.

Septic Tanks. Over 17 million families in the United States use septic tanks or other types of subsurface seepage to dispose of their wastes. This is about 30 percent of the population.

Assuming that about 50 gallons per capita are disposed of in this manner per day, about 3.5 billion gallons of waste are introduced into the soil each day. In addition, there are business and commercial sources that use septic tanks. The wastes discharged to septic tanks contain human wastes, metals from corrosion products, and inorganic and organic chemicals used in the home or business.

Municipal Sludge. In 1980, about 6.8 million dry tons of sewage sludge were produced in the United States [4]. Of this amount about 70 percent was

Table 4–1. Estimates of the Amount of Wastes Applied to the Soil

Source	Amount	Comment
Septic tanks	3.5 to 4 billion gallons per day	contains organics, nutrients, pathogens, metals, and organic chemicals used in the home
Municipal sludge	4.8 million dry tons per year	contains organics, nutrients, and inorganics and organics contributed by homes, industry, and business
Hazardous waste	113 million metric wet tons to surface soils and 67 million metric wet tons disposed of by injection wells	inorganic and organic constituents from industry

disposed of in or on land: 12 percent landspreading on food chain crops, 12 percent landspreading nonfood chain crops, 18 percent distribution and marketing (which ultimately is returned to the soil somewhere), 15 percent landfill, and 12 percent long-term lagooning or other. Thus about 4.8 million dry tons of sewage sludge containing organics and inorganics from homes, business, and industries reach the soil of the nation each year.

Hazardous Wastes. Estimates of the quantities of hazardous waste generated by industry vary depending on the definition of hazardous waste that is used. The Congressional Budget Office [5] has estimated that about 266 million metric tons (MMT) (wet weight) of hazardous waste were generated in 1983 and that about 280 MMT may be produced in 1990. In 1983 about 113 MMT were disposed of using surface impoundments, landfills, and land treatment. All of this material reaches the surface soils. In addition in 1983, about 67 MMT were disposed of by injection wells to deep confining geologic formations.

Agricultural Sources. Agricultural sources contribute fertilizers, pesticides, salts in irrigation water, and organic and inorganics in manures. Manures have been of concern primarily on lands adjacent to high-density animal production operations.

The list of sources that contribute chemicals to the soils can be extended. The quantities that are applied to the terrestrial environment are not trivial and the types of chemicals that are applied are diverse. The quantities that will be applied will increase as the population increases, industry expands, and as other disposal sinks are constrained.

The fact that there have not been more instances of water and soil contamination is testimony to the extensive assimilative capacity of the soil. The capacity must be understood more clearly if greater incidences of contamination are not to occur and portions of this resource are not to be lost.

BASIC CONCEPTS

Soil—The Treatment Medium

This section is not intended to be exhaustive and will only present the major aspects that affect the assimilative capacity of the soils. There are many texts and references that can provide detailed information.

Soil is a composite medium containing inert rock, gravel, and sand, reactive clay minerals, organic matter (living and dead vegetative and animal matter and micro- and macroorganisms) and air and water. It is an entity produced by the interaction of climate, vegetation and fauna which change the basic geologic material.

Soil profiles form from parent material acted on by environmental forces over time. The formation of soil is controlled by:

- temperature and precipitation;
- living organisms and vegetation;
- texture, structure and chemical and mineralogical composition of the parent material;
- topography;
- time the parent materials are subjected to soil formation.

The development of a pH profile in the soil results from the dissolution of alkali salts in the soil and their percolation through the profile. The pH of the soils will influence the solubility, sorption, and movement of constituents applied to the soil.

The structure of a soil results from different particle sizes and aggregates throughout its depth. The surface soils (A_1 horizon) commonly have granular aggregates due to tilling and plant root growth. Such soils may agglomerate when wet. The next lower horizon of soil (A_2 horizon) may have a structure similar to that of the surface soils but also may have platelike horizontal aggregates. Such material may impede air and water improvement.

The lower soils (B and C horizons) contain material more similar to the native geological formation, will have a lower organic fraction, and be less weathered and can restrict water movement.

The pore space of a soil is the portion occupied by air and water. The amount of pore space is determined by the arrangement of the solids. Macropore space allows ready movement of air and percolating water. Movement of water through the soil profile is a function of soil pore space. In the unsaturated zone, the macropores are filled primarily with air. In saturated soils, the pores are filled with water.

The permeability, or hydraulic conductivity of a soil describes the ease of water transmission through the soil. The permeability depends on the particle size, pore space, and bulk density of the soil. The finer the soil texture, the slower the permeability. Knowledge of the soil permeability permits the movement of water and potential pollutants to be estimated.

Each soil has a physical, chemical, and biological environment. As the waste moves through a soil, it contacts solid-liquid and gas-liquid interfaces at which sorption, degradation, and gas transfer can occur. The surface soils consist of inorganic and organic colloids which contribute to the ion exchange and pollutant retention properties of the soil. The organic colloids consist of humus and nonhumus fractions. The nonhumus fractions act as a source of food and energy for the soil organisms and the humus fraction helps maintain good soil structure, increases the cation exchange capacity (CEC) and the organic matter content, and increases the water holding capacity of the soil.

The inorganic colloids are represented by clay minerals. Two groups of clays generally exist: silicate clays characteristic of temperate regions and iron and aluminum oxide clays found in the tropics and semi-tropics.

Microbial activity is a significant aspect of the soil environment and is responsible for the biochemical conversion of waste products in the soil. In a waste amended soil, the type and number of microorganisms depend upon factors such as moisture content, aeration, nutrients, and the type of organic matter. The largest population of microorganisms occurs in the surface soils. In such soils, there are aerobic conditions and ample organic matter. Microbial populations decrease with depth. The soil organisms include bacteria, actinomycetes, fungi, earthworms, and microarthropods.

Comparison to Other Processes

The factors involved in the assimilative capacity of soils are similar to those that occur in other environmental control processes. Perhaps the closest analogy is a municipal or industrial waste treatment plant. In the soil, the applied wastes are degraded, transformed, immobilized, and transported. These processes also occur in conventional municipal and industrial waste treatment systems. A major difference is that with the use of soil to assimilate the wastes, the processes occur in an unconfined reactor filled with soil particles while with conventional systems the processes occur in separate tanks filled with a liquid or slurry. The rates of reaction also are different in the two systems.

A conventional municipal or industrial wastewater treatment system achieving secondary treatment or the equivalent will consist of the following processes: sedimentation, biological treatment, and sludge stabilization and disposal. If needed, tertiary treatment systems can include nutrient removal, adsorption, ion exchange, filtration, and chemical precipitation processes.

All of these processes occur automatically when the soil is used to assimilate wastes (See Figure 4–1). The soil is a biological-physical-chemical reactor that contains: (a) particles (soil) that filter the applied wastewater and transform

Figure 4–1. The Use of Soil as a Waste Treatment Process.

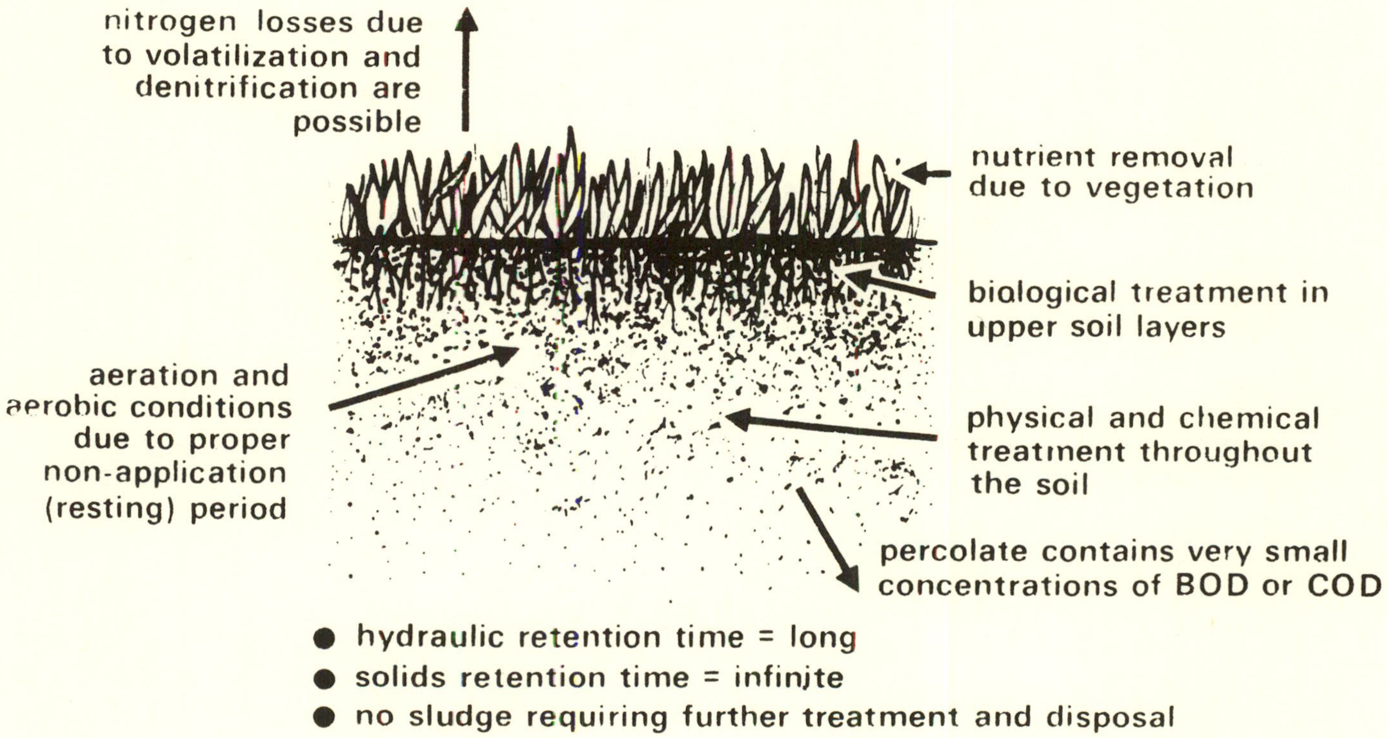

nitrogen losses due to volatilization and denitrification are possible

nutrient removal due to vegetation

biological treatment in upper soil layers

aeration and aerobic conditions due to proper non-application (resting) period

physical and chemical treatment throughout the soil

percolate contains very small concentrations of BOD or COD

- hydraulic retention time = long
- solids retention time = infinite
- no sludge requiring further treatment and disposal

(adsorb, exchange, precipitate) many of the applied chemicals, (b) bacteria and macroorganisms such as earthworms that stabilize the applied organics, and (c) vegetation that utilizes nutrients and inorganics during growth.

Biological treatment occurs as the applied organics are stabilized by the microorganisms in the soil. One important difference is that in the soil, there is no residual sludge that requires subsequent treatment and disposal such as would occur in a conventional treatment system. Any increase in biomass that occurs as the applied organics are stabilized remains in the soil and undergoes natural degradation until it is stabilized and becomes part of the soil humus.

Other differences of land treatment, as compared to conventional wastewater treatment systems, include the mechanism of oxygen transfer, particulate removal, and nutrient control. The predominant oxygen transfer in the soil is gas diffusion and transfer of the oxygen across large surface areas of thin liquid films on the soil particles.

Particulates are removed at the surface or within the soil. Nitrogen removal can occur due to volatilization and to denitrification in the soil as well as by uptake of vegetation that may grow on the site. Phosphorus removal results from chemical precipitation and adsorption reactions in the soil and also by vegetative uptake. Metals can precipitate and adsorb to soil particles and exchange sites in the soil.

Retention time is an important design and operational parameter for waste management systems. In conventional wastewater treatment systems, liquid retention times are short—in the range of 6 to 24 hours—and solids retention times are about 10 days or greater depending upon the treatment and stabilization processes used. In the soil, the liquid retention times are longer, many days to months, depending upon the amount of liquid that is applied and the characteristics of the soil. The solids retention time in the soil can be considered almost infinite since no particles should pass through the soil and few move overland with any surface flow. Because of the longer retention times, the soil is capable of a higher degree of treatment than conventional wastewater treatment systems.

General Reactions

The assimilative capacity of soil at a site is related to the physical, chemical, and biological reactions that take place in the soil. Detailed information on the reactions is presented adequately in other publications [6–11]. Only a limited description of the more important reactions that establish the assimilative capacity of a soil is presented in the following paragraphs.

The source of the waste (municipal, industrial, agricultural) or the form (liquid, semi-solid, or solid) is less important to the soil assimilative capacity than is the type and amount of constituents contained in the waste. The treatment mechanisms and rates of reaction of a chemical (e.g., a metal, an organic) depend on the soil characteristics and environmental conditions and are very much the same regardless of source and type of waste (municipal effluent, sludge, industrial waste, hazardous waste). Each soil has a different assimilative capacity of the waste constituents added to it and organic and inorganic constituents can have different assimilation rates at different sites.

The major reactions involved in the assimilative pathways of a soil are:

- microbiological or chemical degradation of organic compounds;
- immobilization and chemical reaction of metals and cations with little significant migration;
- movement of anions such as nitrate and chloride with the soil water. Phototransformations, volatilization, and vegetative uptake also can occur when wastes are applied to the soil and affect the waste assimilative capacity of a site.

The extent to which these reactions are important in the use of the soil assimilative capacity depends upon the type of land/soil system used to assimilative wastes. For instance, all three reactions are important in land treatment systems. Land treatment is a managed treatment and ultimate disposal process that involves the controlled application of a waste to a soil or soil-vegetation system. The wastes are applied to the surface or mixed with the upper zone [0–1 ft. (0–0.3 m)] of soil. The objective of land treatment is the biological degradation of organic waste constituents and the immobilization of inorganic waste constituents. Controlled migration of certain inorganic constituents such as chloride or nitrate may be permitted if drinking water standards are met. Municipal wastewaters and sludges as well as nonhazardous and hazardous industrial wastes are treated by land treatment processes.

A land treatment system is designed and operated to stay within and not exceed the soil assimilative capacity at a site. Waste applications are not continuous and operating conditions maintain aerobic conditions in the upper soil layers to enhance aerobic microbial degradation of organics.

In contrast, a landfill is a disposal site in which large quantities of organics and inorganics are accumulated. Anaerobic conditions prevail in the landfill and surrounding soil. Suitable liners are used to prevent migration of the contained wastes. Movement of soluble constituents is limited by the liners and by minimizing water movement through the landfill. The characteristics of a landfill can cause the assimilative capacity of the soil immediately surrounding the landfill to be exceeded. The contaminants moving from the landfill are attenuated by the liners and by soil around the landfill.

Underground injection wells offer another example in which the land is used for waste management. Underground injection permanently stores liquid waste in underground strata. There is little chemical or microbiological degradation of organic compounds after the waste is injected and movement of mobile ions will occur only if there is movement of the liquid from the injection zone through the confining strata. Such movement can occur through abandoned wells, cracks, and fissures. The subsurface soil assimilative capacity of underground injection is not considered when this technology is used except in an attempt to ensure confinement of the injected liquid.

Limiting Constituent Analysis

The determination of the soil assimilative capacity of a site used for waste treatment or disposal is important to (a) protect human health and the environment and (b) determine the feasibility of the land as a waste management alternative. The assimilative capacity determination considers: (a) the waste constituents of greatest potential environmental concern, (b) the transformation and immobilization reactions that will occur in the soil, and (c) the required management of the site. The assimilative capacity of the site can be determined using: (a) known physical, chemical, and biological reactions, (b) sound scientific and engineering principles, and (c) practical field experience.

The limiting constituent analysis integrates the waste characterization and site assimilative determinations and provides a focus on those constituents that are of greatest concern [12]. The limiting constituent or limiting parameter concept is based upon the fact that soil has an assimilative capacity for inorganic and organic constituents. The constituent that results in the lowest application rate to avoid health and environmental problems is the limiting constituent. Use of this application rate means that the other waste constituents are applied at a conservative rate and should not reach levels of environmental or health concern.

A variety of parameters can limit waste application rates. Examples include: (a) nitrate leached from the site to groundwater, (b) cadmium in food chain crops, (c) synthetic organic compounds in surface, groundwater, and crops, (d) salts that inhibit seed germination, or (e) metals that may be toxic to crops. Figure 4–2 illustrates the use of the limiting constituent concept with an industrial waste for which land treatment was being considered. In this example, one or more of the synthetic organics were the limiting constituents requiring the largest land area to void exceeding one of the concerns: i.e., water quality criteria, rate of decomposition, or food chain impacts. One of the metals was the next limiting constituent if the constraint due to the synthetic organics was relieved by reducing the concentration in the applied waste.

There are three principal results that occur from the limiting constituent analysis. The first is a determination of the application rate which does not exceed the assimilative capacity of the soil at a site. The second is a determination of which sites are acceptable for the land treatment of the waste under consideration. The third is the identification of critical constituents that control or limit the recommended application rate. The first two results are important for land acquisition and subsequent engineering design. The third is important because the land limiting constituents determine:

Figure 4–2. An Illustration of the Use of the Limiting Constituent Concept for the Land Treatment of an Industrial Waste.

WASTE CONSTITUENT	ITEM LIMITING APPLICATION RATE	ESTIMATED LAND AREA NEEDED
Nitrogen	-nitrate in groundwater	
Synthetic organics	-surface water quality criteria -drinking water standards -food chain impact	
Metals	-food chain impact -water quality criteria or standards	
Water	-soil permeability	
Biodegradable organics	-maintenance of aerobic conditions -surface clogging of soil	

1. the most appropriate or critical parameters and locations (plants, soils, groundwater, etc.) for monitoring;
2. the specific constituents for which industrial in-plant source control or pretreatment would lead to a direct reduction in land treatment systems costs; and
3. the most relevant parameters on which to focus in order to improve waste characterization or site assimilative capacity information.

TRANSFORMATIONS AND TRANSPORT

Introduction

The ability to predict the transformations, transport, and fate of constituents added to the soil is key to important decisions such as:

1. maintenance of environmental quality, human health, and the quality of the soil resource for food production;
2. site selection for waste disposal options;
3. exposure assessment and risk management;
4. remedial action should adverse conditions occur;
5. determination of the impact of new chemicals being released into the environment.

Public and governmental agencies continually are asked to make judgements on these and related matters. Results from laboratory and field studies are needed to provide the basis for such decisions.

Diverse constituents are in wastes that are treated and disposed of in soil. The transformations and transport of some of the constituents have been studied for decades and are reasonably well known. For other constituents, especially synthetic organics, there is less information.

Most transport and fate studies in the soil have been concerned with inorganic contaminants (salts and metals) with the result that considerable information exists on these contaminants. There also is considerable information on the transport and fate of nitrogen and phosphorous in the subsurface.

In contrast, current information permits only generalized conclusions concerning the movement of a few synthetic organics in the subsurface environment. Mathematical models capable of providing a first generation predictive ability for selected organics are under development in several locations.

Adequate knowledge about the soil assimilative capacity is available to avoid gross health and environmental problems associated with conventional pollutants (oxygen demanding compounds and solids), many nonconventional pollutants (oil and grease, metals, salts) and nutrients (nitrogen and phosphorous). This knowledge is not always used well when wastes are applied to the soil and situations where the assimilative capacity is exceeded and the soil or groundwater is contaminated continue to occur.

In addition to understanding the transport and fate of contaminants when added to the soil, it also is important to understand and adequately characterize the conditions in subsurface soils. For instance, subsurface soils below the root zone have been considered as having limited microbial content and activity. Newly developed techniques permit the direct enumeration and evaluation of microorganisms in the subsurface soils. Microbial concentrations of from 3 to 5 million live microorganisms per gram of dry material have been found at 5 meters depth in several soils. The organisms found at this depth were able to slowly degrade organics such as toluene, chlorobenzene, trichloroethylene, and tetrachloroethylene [13].

The following sections identify the knowledge associated with the transformation of specific waste constituent categories and the possible transport of the constituents. Also included are comments about the state of knowledge and additional knowledge that is needed.

Carbon

The carbon in wastes is predominantly organic carbon in the residues of plants and animals, i.e., nonsynthetic carbon. When wastes are added to the soil, the more easily degradable materials are metabolized rapidly with cellulose, hemicellulose, lignins, and some synthetic organics persisting for longer periods of time.

The soil has a large capacity to remove organic matter from wastes. The organic carbon added in a waste ultimately will exist as carbon in the protoplasm of soil microorganisms and in plant tissue, as slowly degradable soil humus, and as carbon dioxide released to the atmosphere. None of these compounds has a direct adverse environmental or health effect. Biodegradable organic carbon is not an element of environmental quality concern.

Nitrogen

Nitrogen will continue to be a potential contaminant of concern in land-based waste management systems, primarily because of the mobility of nitrate nitrogen and the low concentration that can be tolerated in potable water supplies. Humid areas with sandy soils and concentrations of humans or animals and with intensive use of fertilizers will be where the soil assimilative capacity is likely to be exceeded and groundwater contamination occurs.

Nitrogen can be mineralized, immobilized, nitrified, denitrified, and volatilized. Mineralization is the process in which nitrogen is converted to a form that is both mobile in the soil-water system and available to plants. Organic nitrogen is converted to ammonium nitrogen which is oxidized (nitrification) by aerobic microorganisms in the soil to nitrite and nitrate nitrogen. Immobilization occurs when the nitrogen is tied up as organic nitrogen such as in microbial cells. In the soil the processes of mineralization and immobilization occur simultaneously. Under anaerobic conditions, the oxidized nitrogen can be reduced (denitrification) by facultative microorganisms to gases such as nitrous oxide and molecular nitrogen.

Ammonium ions are positively charged and move slowly with the soil water because of the attractive forces between the ammonium ions and negatively charged clay and organic colloids. As long as the nitrogen stays in the ammonium form, the possibility of nitrogen loss by leaching is low. However, in aerobic soils ammonium nitrogen is oxidized to nitrate by microorganisms. Nitrate leaching can be of concern if large amounts of nitrate are present when irrigation or rainfall exceeds soil water storage or crop moisture requirements. Under such conditions, excess water and soluble nitrates will move through the soil. The nitrification rate in soil is affected by soil aeration, temperature, and soil moisture.

Once the nitrates are below the root zone, little opportunity exists for utilization by a crop or for denitrification. Denitrification can be an important method of nitrogen loss from soils and occurs when nitrates are present and when there is poor aeration and a high oxygen demand.

The waste application rate to soil frequently is limited by the amount of nitrogen in the wastes and the assimilative capacity of the site for nitrogen. The nitrogen application rates should be consistent with nitrogen needs and crop utilization rates so that the amount of excess nitrogen is minimized.

State of Knowledge

Concerns over nitrate contamination of groundwater and optimum use of the fertilizer nutrients in wastes are of long standing. Research on these topics has been supported by the Department of Agriculture (USDA), the Environmental Protection Agency (EPA), the National Science Foundation (NSF), land-grant institutions, and industry. This research has been conducted to determine proper application rates of animal manures, wastewater treatment sludges, wastewater and fertilizers. Much is known about the fundamental mechanisms of nitrogen transformation and transport and about how to avoid exceeding the assimilative capacity of a soil for nitrogen. This knowledge has been translated into design and operating guidelines that can be used when wastes and residues are applied to the land. These guidelines appear satisfactory to prevent gross nitrate contamination of groundwater when wastes are applied to land.

The research needs that can improve our understanding of nitrogen cycling and losses in soils, strengthen the existing guidelines, and further decrease environmental and health problems are:

- refine the knowledge of nitrogen behavior in soils that receive waste, especially with regard to ammonia volatilization, nitrogen mineralization, nitrification, and denitrification;

- determine nitrogen cycling in forest lands and determine the assimilative capacity of forest soils so that valid criteria can be developed for waste utilization on such lands;
- develop management techniques for nitrogen removal under field conditions when wastes are applied to soils.

Metals

Wastes can add metals as well as organic matter and nutrients to a soil. If in excess, metals such as cadmium, copper, nickel, selenium, and zinc have the greatest potential to contaminate the animal and human food chain. The soil does have an assimilative capacity for metals which can be used to minimize environmental and health problems. Federal and state regulations and guidelines have been developed to reduce the probability of such problems.

The soil-plant barrier [14] helps protect the plant-animal-human food chain from toxic metals. Cadmium, selenium, and molybdenum are metals which can escape the soil-plant barrier while beryllium and cobalt are metals which may escape this barrier. The soil-plant barrier can be circumvented when animals directly ingest soil, sludge, or vegetation to which soil or sludge particles adhere.

Most metals, with the exception of boron, are strongly adsorbed by soil particles and their concentrations in soil solutions are low. Adsorption by clay minerals, metal oxides, and organic matter is the predominant mechanism of metal immobilization in soils. Soil pH is the main factor controlling soil adsorption-desorption processes and metal solubility. With the exception of molybdenum and selenium, all of the metals are more soluble at low pH values.

Where wastes are applied to soils, the soil pH should be maintained at 6.5 or greater to minimize mobility of the metals. For most soils, except acid soils strongly buffered by aluminum, a pH of 6.5 can be maintained by periodic addition of lime.

Many states have limits on the total cumulative amounts of cadmium, copper, lead, nickel, and zinc that can be applied to cropland. Such limits are based on the assimilative capacity of the soil and control the number of years that a site can be used for waste application as well as the total amount of metal and waste that can be applied to the site.

The cation exchange capacity (CEC) of the soil commonly is used to estimate the ability of a soil to minimize metal uptake by plants. The CEC is a measure of the net negative charge associated with clay minerals and organic matter in the soil. The negative charge on soil organic matter and to some extent on clay minerals is pH-dependent; an increase in pH increases the negative charge.

Cumulative metal recommendations based on the CEC of a soil have been developed by researchers in various land grant universities, USDA, and EPA. CEC is used as a soil property that is measured easily and is related to soil components that minimize the plant availability of metals added to soils. It is a general but imperfect indicator of soil components such as organic matter, clay, and metal oxides that limit the solubility of most metals. If the soil pH is maintained at a pH of 6.5 or greater *and* if the application of waste to a site ceases before the cumulative limits are exceeded, the soil assimilative capacity should not be exceeded and the soil should be able to be used for the growth of any crop without adverse effects on crop yield or the food chain.

State of Knowledge

In the last fifteen years, extensive research has been conducted on the transformations and fate of metals added to soils. The considerable information that is available appears adequate to avoid exceeding the assimilative capacity of soils. Available information indicates that when wastes are applied to land according to current federal, state, and provincial guidelines, it is unlikely that there will be deleterious environmental or health problems.

The research needs that can improve our understanding of the assimilative capacity of soils for metals include:

- increased evaluation of the availability of waste applied metals in soil to plants and animals;
- identification of soil parameters other than CEC to establish annual and cumulative limits of waste applied metals.

Pathogens

The removal of microorganisms from wastes as they contact the soil is an important environmental and health consideration. Bacteria and viruses are removed from wastewater as it percolates through the soil. Unless fissures exist in the soil, bacteria and viruses are removed in a short distance. Most of the organisms in the applied waste are removed, generally greater than 90 percent removal, in the surface layer of soil. Several feet or more of soil are necessary for near complete removal of the applied microorganisms.

Removal of bacteria, viruses, protozoa, and helminths (worms) in applied wastes occurs by filtration, adsorption, desiccation, predation, competition for food, and exposure to other adverse conditions. Bacteria and viruses can move through the soil if the soil is porous, such as at a rapid infiltration land treatment site, and where there is a high water table.

Municipal wastes and the sanitary wastes from industry will contain coliform organisms, bacterial pathogens, helminthic parasites, and viruses. Only waste treatment sludges treated by a process to "significantly reduce pathogens" should be applied to land used to grow human food chain crops. Such processes include anaerobic or aerobic digestion, air drying, composting, and lime stabilization.

Table 4–2. Organic Chemicals Commonly Detected in Public Water System Wells*

Bis (2–ethylhexyl) phthalate
Toluene
Di-n-butyl phthalate
Trichloroethylene
Ethylbenzene
Diethyl phthalate
Benzene
Anthracene/phenanthrene
Butyl benzyl phthalate
Trichlorofluoromethane

*Adapted from Reference 3, Table 4.2

State of Knowledge

Although concerns have been expressed about the addition of human and animal pathogens when wastes are applied to soils, there have been no serious disease problems caused by the application of wastewaters or stabilized sludges on agricultural cropland when proper guidelines and permit conditions are followed. Soils that are suitable for waste disposal from the hydraulic, nutrient control, and adsorptive standpoint can provide satisfactory removal of pathogenic microorganisms.

A research need that can improve our knowledge of pathogen removals in soils and can decrease the concerns of the public is: conduct monitoring studies on the survival and transport of waste applied viruses under a wide range of field conditions.

Synthetic Organics

Organic contaminants in groundwater are an increasing problem. Synthetic organic chemicals enter the soil as a result of (a) the use of such chemicals in agriculture (pesticides), (b) landfills used for the disposal of organic compounds, (c) leakage from waste storage ponds and underground storage tanks, (d) spills, and (e) uncontrolled waste disposal sites. For many of the synthetic organics now in use and that may reach the soil, little is known about their toxicity, degradation, and mobility. With the improvements in analytical chemistry, instances of organic contamination of groundwater are more common. Organic chemicals commonly detected in public water supply wells are noted in Table 4–2.

Biological degradation, volatilization, adsorption, chemical degradation, phototransformation, and possibly vegetative uptake can prevent or retard the movement of synthetic organics from surface soils to the groundwater. The synthetic organics that pose the greatest concern to groundwater and potable water supplies are those that are relatively soluble, nonvolatile, and nonbiodegradable.

The fact that numerous compounds are found in groundwaters in concentrations of potential concern indicates that the assimilative capacity of the soil has been exceeded for such compounds. Specific information on the rates of degradation, adsorption, and volatilization of synthetic organic chemicals is needed to establish site specific soil assimilative capacities and the rates at which such chemicals can be applied to the soil.

In determining the site assimilative capacity for a chemical, it is important to know (a) how far the chemical will move under the site-specific climatic and soil conditions and (b) how much of the applied chemical will remain by the time it moves past the bottom of the treatment zone or it reaches the groundwater aquifer. In addition it is importnat to know the concentrations of the chemical in the groundwater so the health and environmental risks can be estimated.

Preliminary estimates concerning the mobility of individual compounds can be made by comparing the time required to degrade a compound (t_D) in the soil to the time it takes for the compound to migrate out of the biologically active treatment zone (t_m). This is a conservative comparison since volatilization and other losses are not included and it is assumed that no degradation or adsorption occurs below the biologically active treatment zone. This preliminary screening approach can identify those compounds that may migrate beyond the treatment zone and that should be subject to detailed evaluation. For such compounds the effect of volatilization, possible plant uptake and additional degradation can be considered using potentially applicable models [15, 16].

In the preliminary estimate, if t_D is less than t_m, the compound will degrade before it moves from this defined treatment zone. Compounds with t_m values greater than t_D are mobile and persistent and are those of greatest concern.

The water movement through the soil needs to be known or estimated from (a) climatic records and the amount of liquid or wastewater applied at the site and (b) the permeability of the soil. With the water movement estimated, t_D can be estimated if the rate of degradation (K_D) or degradative half-life of the compound is known or estimated. The migration time, t_m, can be estimated if the sorption rate or coefficient (K_s) is known or estimated. Information on the half-life of some organic compounds is available [9, 17].

Reported values for the half-life of a compound may vary by an order of magnitude especially between field and laboratory data. Although the information is important, it is difficult to obtain and assess. Identified ''degradation losses'' commonly include other pathways of loss and are affected by temperature, moisture content, acclimation of the microbial population, and the existence of aerobic or anaerobic conditions.

Because of the difficulties of determining the actual degradation half-life of a compound as well as difficulties of determining other loss rates such as those due to volatilization, chemical degradation, precipitation, and sorption, an overall loss rate or half-life frequently is considered when identifying the transformations and transport of a synthetic chemical in the soil. These overall loss rates are site and chemical specific and are affected by local environmental considerations. Such overall loss rates and constants are mathematically simple and operationally useful. However, such rates and constants must be recognized for what they are, the equivalent of "black boxes" that do not permit an understanding of the phenomena that are involved, the predominance of specific loss mechanisms, or the importance of the factors that affect the loss rates.

Such rates and constants also do not permit an extrapolation of loss rate information to other situations. Information about specific loss rates and constants and how such rates and constants are affected by chemical concentrations and environmental factors is needed to understand the transformations and transport of synthetic organics added to soils.

Sorption is an important factor that determines the movement of chemicals in the soil. Adsorption and desorption retard the migration of chemicals in soil. Knowledge of these processes is important in understanding chemical transport in soil and the assimilative capacity of soils.

Adsorption and desorption usually are considered together as a reversible process. Both the Langmuir and Freundlich isotherm models have been used to describe the sorption process. The Freundlich isotherm is an empirical model expressed as:

$$C_s = K_s \cdot C^{\frac{1}{n}} \qquad (4.1)$$

where C_s = concentration of chemical adsorbed on soil (g/kg), K_s = adsorption (partitioning) coefficient (m^3/kg), C = concentration of chemical dissolved in the soil moisture and n = Freundlich constant.

At low concentrations, the adsorption of many organic chemicals is proportional to the concentration, n = 1, and the shape of the isotherm is approximated by a straight line [15]. In this case:

$$C_s = K_s \cdot C \qquad (4.2)$$

For organic chemicals, adsorption is related to the soil organic matter and this parameter can be used to normalize adsorption partitioning coefficients. The constant K_{oc} (the adsorption coefficient on organic carbon) can be used in equation 4.2 instead of K_s. These coefficients are related:

$$K_s = \frac{K_{oc}\ (\%\ OC)}{100} \qquad (4.3)$$

where % OC is the organic carbon content of the soil as a percentage.

For most soils, the organic matter (% OM) rather than the organic carbon

Table 4–3. Correlations of K_{oc} and K_{ow} Values

Compounds	Correlations	Reference
s-triazines and dinitro analine herbicides	$\log K_{oc} = 0.94 \log K_{ow} + 0.02$	[18]
polycyclic aromatic hydrocarbons	$\log K_{oc} = 1.0 \log K_{ow} - 0.21$	[18]
thirty pesticides	$\ln K_{oc} = \ln K_{ow} - 0.73$	[22]
	$\log K_{oc} = 0.72 \log K_{ow} + 0.49$	[23]
	$\log K_{oc} = 2.0 \log K_{ow} - 0.317$	[23]

content is known. The % OC can be estimated if the carbon content of the organic matter is known:

$$\% \text{ OC} = (\% \text{ OM}) \cdot F_{oc} \tag{4.4}$$

where F_{oc} = the fraction of the organic matter that is carbon.

K_{oc} values also are needed to permit an estimation of the sorption coefficient K_s. It has been suggested [18] that the role of soil organic matter in adsorption of neutral organic molecules is similar to that of an organic solvent in liquid-liquid extraction. A large data base on octanol-water partition coefficients (K_{ow}) exists for many pesticides and synthetic organics and additional data on the K_{ow} values for other organics is becoming available [19–21].

Correlations between K_{oc} and K_{ow} values have been developed and it appears that the two values are closely related (See Table 4–3). The advantage of such correlations is that once one of the parameters is known, the other parameter can be estimated. For specific synthetic organic chemicals, K_{oc} can be estimated from K_{ow} values and K_s can be determined using K_{oc} values and the organic carbon fraction of a soil. Thus, from a limited amount of information about a soil and about chemical partitioning and degradation coefficients, simple models can estimate the expected distribution of a chemical in a soil and the likelihood that groundwater contamination will occur.

The K_{ow} and K_{oc} relationships have been determined for soils that have an organic carbon greater than 0.1 percent on a weight basis. However, many permeable soils have an organic carbon content less than 0.1 percent. The water movement is high in such soils due to their permeability and the potential for groundwater contamination by soluble organics also is high due to the low organic carbon content.

As indicated with the discussion of equation (4.2), a linear sorption isotherm commonly is assumed. Some synthetic organic compounds may have nonlinear isotherms, especially where high concentrations exist in soils, such as at the boundaries of landfills and in intensively used industrial waste land treatment systems.

To critically identify the soil assimilative capacity of synthetic organic chemicals, research is needed to (a) determine the partition and degradation coefficients of specific organics; (b) verify the assumption of linear isotherms and the effect of low organic carbon soils; (c) develop models that adequately take into account the various loss mechanisms and the stochastic nature of water moving through the soil; and (d) verify the models under field conditions.

Equally important is research to delineate the transport that occurs when the chemicals exist as a mixture. Available field information suggests that, when present in a mixture with other organics, an organic chemical can move more rapidly through the soil than predicted from theory or from information based upon single compound rates and information. The phenomenon of "facilitated" transport needs further elucidation since, under practical conditions, organic chemicals will exist as mixtures when they are added to the soil.

Leachate processes generally are emphasized when considering the transformation and fate of wastes added to the soil. Volatilization commonly is neglected or assumed to be insignificant. The factors affecting volatilization and gas transport of chemicals in soil are poorly understood and the ability to predict such rates of loss is very poor. Such knowledge is needed to: (a) understand this loss mechanism, (b) develop sound technical approaches using the soil assimilative capacity, and (c) avoid multimedia transport of pollutants.

State of Knowledge

Less is known about the assimilative capacity of soils for synthetic organics than for any other potential contaminant. At the same time, an increasing array of synthetic organics of potential environmental and health concern are being applied to soil for various reasons and many are found in groundwater.

To increase our knowledge about the assimilative capacity of soils for synthetic organics and thereby minimize environmental and health risks, the following research needs to be undertaken:

- obtain a better understanding of the number and type of microorganisms in the deeper subsurface soils and their role in the removal of contaminants in soil water and groundwater;
- determine the degradation rates and half-lives of synthetic organics in soils under varying environmental conditions;
- determine the partition (sorption) coefficients of synthetic organics in soils having different clay and organic contents;
- verify and, if needed, extend the apparent relationship between K_{ow}, K_{oc}, and K_s for different synthetic organics and different soils;
- determine the bioconcentration of synthetic organics applied to the soil by soil invertebrates, by plants, and by animals;
- develop a data base for synthetic organics which includes information on K_{oc}, K_{ow}, K_s, physical properties, volatilization rates, plant uptake, bioconcentration factors, and half-lives in soils;

- determine the extent to which the linear isotherm assumption is valid for soils in the field;
- investigate the phenomena of ''facilitated'' transport of organic chemicals in soils and determine the factors that cause such transport;
- determine the rates of volatilization of synthetic organics when wastes are applied to land and the importance of this loss mechanism;
- develop protocols that can estimate the treatability of synthetic organics in wastes that are applied to the soil;
- develop screening and definitive models that can be used to determine the synthetic compounds of environmental concern when added to the soil and the soil assimilative capacity for such compounds.

RESEARCH RECOMMENDATIONS

Research needs required to fill the gaps in our knowledge about the soil assimilative capacity for wastes have been identified in the previous sections. In addition, there are research needs that relate to management and the overall understanding of the soil assimilative capacity.

Information concerning the soil assimilative capacity comes from a number of disciplines and requires multidisciplinary research efforts. In some cases, individuals with a traditional disciplinary training can successfully undertake the research and develop the insightful interpretations of data and theory. However, experience indicates that individuals having multidisciplinary backgrounds and research experience can make greater contributions. Because the questions related to the soil assimilative capacity are not routine, individuals having such a ''hybrid'' background are needed for the identified research.

Research Needs

For the reasons put forth in the previous section, the priority research concerning the soil assimilative capacity relates to synthetic organics. A number of pertinent research needs were identified in that section. These can be grouped under general categories:

1. determine the rates of degradation, immobilization, volatilization, and transport of synthetic organics when such compounds are added to soils;
2. develop screening and predictive models or protocols to determine the assimilative capacity of soils for synthetic organics of concern;
3. identify waste application rates that will permit the soil assimilative capacity to be used for the treatment and disposal of wastes without causing environmental and health problems.

In addition to the needs related to synthetic organics, there are other important research needs that can enhance our knowledge of the soil assimilative capacity. These include:

1. Field verification of predictive design. Much can be learned from field investigations (essentially post mortems) of existing sites, especially those at which groundwater contamination or other environmental problems have occurred. These sites commonly were established using the then existing good judgement, information, and design. Some flaw in the assumptions or the design or modifications in operation may have occurred. Knowledge of the causes of previous environmental problems will improve future use of the soil as a waste management alternative, better protect the land/soil resource, and allow the soil assimilative capacity to be better understood.
2. Determine operational practices that can be used to maximize and extend the use of land treatment and disposal sites. Such sites can receive different wastes over a period of time. The assumptions and rate constants used to estimate the initial site assimilative capacity may not be adequate over time or with different wastes. Methods that can be used at a site to avoid environmental and health problems when conditions change or assumptions prove invalid should be determined.
3. Develop monitoring protocols and practices that can assure that the soil assimilative capability is not exceeded. Land treatment and disposal sites are dynamic systems. Adequate monitoring procedures are needed to identify when the soil assimilative capacity begins to be exceeded. The type and location of monitoring equipment, the parameters to be monitored, and the frequency of monitoring need to be determined.

The research described in the above paragraphs is of a long-term nature, is both fundamental and applied, and can be done in the laboratory or the field. Both NSF and EPA are appropriate sources for support of such research.

The results of many of the research projects will have an impact on the guidelines, regulations, and controls that may have to be developed to protect human health and the environment and to avoid exceeding the assimilative capacity of the soil. These results will help regulatory decision makers in EPA and thus support of such research by EPA is appropriate.

The Department of Agriculture also should be interested in supporting some of the needed research since USDA has responsibilities to protect the land/soil resource and avoid situations that adversely impact agriculture and the public. Such research can be conducted directly by USDA researchers, through the land grant agricultural experiment stations, or by competitive grants.

Personnel Needs

As indicated earlier, multidisciplinary efforts will be needed to successfully conduct meaningful and insightful research. The disciplines that can conduct such research include chemistry, soil science, environmental engineering, hydrology, and crop science. The research support should not only be for the completion of specific research tasks. The support also should be used to encourage multidisciplinary research efforts and the education and training of individuals who can bridge several of the disciplines and be the "hybrid" investigators needed for future success on these related problems.

The success and development that has occurred in the past decade has resulted from individuals who have been able to cross disciplinary boundaries and apply theory and concepts developed in one area to problems in another. Such efforts should be continued and students should be educated in a manner that encourages such cross-disciplinary thinking and research.

REFERENCES

1. National Research Council, *Groundwater Contamination,* Geophysics Study Committee, National Academy Press, Washington, D.C., 1984.
2. Pye, V. I., R. Patrick, and J. Quarles, *Groundwater Contamination in the United States,* University of Pennsylvania Press, Philadelphia, 1983.
3. Wood, E. F., R. A. Ferrara, W. G. Gray, and G. F. Pinder, *Groundwater Contamination from Hazardous Wastes,* Prentice Hall, Englewood Cliffs, NJ, 1984.
4. Environmental Protection Agency, "Process Design Manual—Land Application Of Municipal Sludge," EPA–625/1–83–016, Municipal Environmental Research Laboratory, Cincinnati, OH, October 1983.
5. Congressional Budget Office, "Hazardous Waste Management: Recent Changes and Policy Alternatives," Congress of the United States, U.S. Government Printing Office, Washington, D.C. 1985.
6. Brown, K. W., G. B. Evans, and B. D. Frentrop, *Hazardous Waste Land Treatment,* Butterworth Publishers, Woburn, MA, 1983.
7. Iskandar, I. K., (ed.), *Modeling Wastewater Renovation—Land Treatment,* John Wiley and Sons, New York, NY, 1981.
8. Loehr, R. C., W. J. Jewell, J. D. Novak, W. W. Clarkson, and G. S. Friedman, *Land Application of Wastes,* Van Nostrand Reinhold, New York, 1979.
9. Overcash, M. R., *Design of Land Treatment for Industrial Wastes, Theory and Practice,* Ann Arbor Science Publishers, Ann Arbor, MI, 1979.
10. Overcash, M. R., *Decomposition of Toxic and Nontoxic Organic Compounds in Soils,* Ann Arbor Science Publishers, Ann Arbor, MI, 1981.
11. U.S. Environmental Protection Agency, *Process Design Manual, Land Treatment of Municipal Wastewater,* EPA 625/1–81–013, Cincinnati, OH, 1981.
12. Loehr, R. C. and Overcash, M. R., "Land Treatment of Wastes: Concepts and General Design," *Journal of Environmental Engineering, 111,* 141–60, ASCE, April 1985.
13. Wilson, J. T., McNabb, J. F., Wilson, B. H., and Noonan, M. J., "Biotransformation of Selected Organic Pollutants in Ground Water," in *Developments in Industrial Microbiology,* Society for Industrial Microbiology, *24,* 225–33, 1983.
14. Logan, T. J. and R. L. Chaney, "Utilization of Municipal Wastewater and Sludges

on Land—Metals,'' in *Utilization of Municipal Wastewater and Sludge on Land,* edited by A. L. Page et al., 235–23, University of California, Riverside, CA, 1983.

15. Bonazountas, M., ''Soil and Groundwater Fate Modeling,'' in *Fate of Chemicals in the Environment,* edited by R. L. Swann and A. Eschenroeder, 41–65, American Chemical Society, ACS, Symposium Series 225, Washington, D.C., 1983.
16. Jury, W. A., W. F. Spencer, and W. J. Farmer, ''Behavior Assessment Model for Trace Organics in Soils: I. Model Description,'' *J. Environ. Qual.* 12:558–64, 1983.
17. Sims, R. C., *Land Treatment of Polynuclear Aromatic Compounds,* Ph.D. diss., North Carolina State University, 1982.
18. Rao, P. S. C. and J. M. Davidson, ''Estimation of Pesticide Retention and Transformation Parameters Required in Nonpoint Source Pollution Models,'' in *Environmental Impact of Nonpoint Source Pollution,* edited by M. R. Overcash and J. M. Davidson, 23–67, Ann Arbor Science Publ., Ann Arbor, MI, 1981.
19. Chiou, C. T., V. H. Freed, D. W. Schmedding, and R. L. Kohnert, ''Partition Coefficient and Bioaccumulation of Selected Organic Chemials,'' *Env. Sci. and Tech.* 11:475–78, 1977.
20. Banerjee, S., S. H. Yalkowsky, and S. C. Valvani, ''Water Solubility and Octanol/Water Partition Coefficients of Organics. Limitations of the Solubility—Partition Coefficient Correlation,'' *Env. Sci. and Tech.* 14:1,227–229, 1980.
21. Rapaport, R. A. and S. J. Eisenreich, ''Chromatographic Determination of Octanol-Water Partition Coefficients (K_{ow}) for 58 Polychlorinated Biphenyl Cogeners,'' *Env. Sci. and Tech.* 18:163–70, 1984.
22. McCall, P. J., R. L. Swann, and D. A. Laskowski, ''Partition Models for Equilibrium Distribution of Chemicals in Environmental Compartments,'' in *Fate of Chemicals in the Environment,* edited by R. L. Swann and A. Eschenroeder, 105–23, American Chemical Society, ACS Symposium Series, 225, Washington, D.C., 1983.
23. Cherry, J. A., R. W. Gillham, and J. F. Barker, ''Contaminants in Groundwater: Chemical Processes,'' in *Groundwater Contamination,* 46–64, National Research Council, Geophysics Study Committee, National Academy Press, Washington, D.C., 1984.

5

GLOBAL AIR CHEMISTRY AND CONTINENTAL-SCALE AIR POLLUTION: AN ASSESSEMENT OF LONG-TERM RESEARCH NEEDS

Steven C. Wofsy

PREAMBLE

Human beings live on a planet characterized by change. Change is evident on all space and time scales, from the very large and very long—the hundreds of millions of years associated with rearrangement of the continents—to the very short and relatively local—the day-to-day variations of weather. The 20th century is a unique time in earth history when one species, humanity, has developed the ability to alter the environment on the largest (i.e., global) scale and to do so within the lifetime of an individual species member. These changes may affect the habitability of the earth: the ability of the planet to support communities of plants and animals, to produce adequate supplies of food, and to sustain and renew the quality of air and water and the integrity of the chemical cycles essential for life.

Humanity has the ability to manage its resources, to plan intelligently for its future, and to preserve the necessary elements of its habitat. On the other hand, if the human race is to be successful in this endeavor it must take steps now to develop the body of knowledge required to permit wise policy choices in the future. The task is urgent. (Adapted from *Global Change: Impacts on Habitability*)

INTRODUCTION

This chapter discusses emerging issues in atmospheric science as related to human impact on the chemistry of the atmosphere. The focus is mainly on direct and indirect effects of industrial and agricultural development. Scientists have only recently become aware that the atmosphere is undergoing slow but profound changes in composition and structure, giving rise to the concern stated so eloquently by the authors of the report cited in the preamble above. In order to understand the causes and the implications of these changes, we must unravel the connections and feedbacks between the atmosphere, the biosphere, the hydrosphere, and the sphere of human activities.

We do not deal at length here with a number of important atmospheric problems for which there are well-developed programs of research, such as the global greenhouse effect due to buildup of atmospheric CO_2. Problems relating to urban air pollution are also not considered. The omissions reflect the author's perception that these important problems are widely appreciated in government and that research priorities are currently being addressed by national programs, with extensive input from scientists at all levels. We consider here primarily changes in the concentration of chemically important species in the atmosphere: CH_4, N_2O, CO, O_3, oxides of nitrogen and hydrocarbons. The spatial scales of interest range from global to continental. In this context, the "acid rain" issue is examined to see if improved scientific return could be obtained by recognition of the intimate relationships between acidity and other pollutants transported on continental scales, such as oxidants and oxidant precursors. The atmospheric chemistry of acid precursors is closely tied to other pollutants, and it is at present impossible to distinguish the deleterious effects of acids from possible impacts due to a variety of other pollutants.

We present first a brief summary of present knowledge related to global and regional tropospheric chemistry and to stratospheric chemistry. We then discuss in general terms public perception of the problem, existing programs and support, and possible approaches to long-term research.

SCIENTIFIC BACKGROUND

Tropospheric Chemistry—Global

Table 5–1 summarizes the composition of the earth's atmosphere, for gases with concentrations greater than 0.1 parts per million by volume (ppmv), and shows also selected components with concentrations less than 0.1 ppm. The major gases (N_2, O_2, Ar, H_2O, CO_2) were discovered in the nineteenth century and their concentrations were measured accurately by manometric (pressure measuring) techniques. The most abundant trace gases (O_3, CH_4, N_2O, CO, H_2) were dis-

Table 5–1 Atmospheric Composition

Gas	Approx. Lifetime (Years)	Abundance		(excluding noble gases) Discovery
N_2	$>10^8$	78%		
O_2	10^7	21%		
Ar	10^9	1%		19th century
H_2O	0.1	0–3%	a,b	
CO_2	10^3	0.03%	a,b,i	
O_3	0.1	0.01–10 ppm	a,i,?	1930 Chapman
CH_4	10	1.6 ppm	a,b,i	1948 Migeotte
N_2O	150	0.3 ppm	a,b,i	1939 Adel
CO	0.1	0.05–0.2 ppm	a,b,i,?	1949 Migeotte (telluric lines in solar spectrum)
NO,NO_2		0.1–10 ppb	a,b,i,?	
C_2H_6		1 ppb		
C_2H_4		0.1 ppb		
C_2H_2		1 ppb	a,i,?	
PAN		?		
HNO_3		0.01–1 ppb		
HNO_4 (peroxynitric acid)		?		
Others				
$CFCl_3$		0.2 ppb		
CF_2Cl_2		0.3 ppb		
CH_3CCl_3		0.2 ppb		
CH_3Cl		0.6 ppb		
OH		?		
HO_2		?		
H_2O_2		?		

Notes:
a—affected by human activity; b—biogenic; i—increasing atmospheric burden; ?—uncertain
Source: Compiled by author.

covered and measured after 1930 primarily by scientists making spectroscopic observations of sunlight passing through the atmosphere. Measurements of atmospheric composition are today made mainly by sensitive gas chromatographic techniques, although there is much promise in emerging laser techniques. Studies of atmospheric composition would be impossible without these modern analytical methods, particularly where minor trace species are involved.

The major gases are chemically stable and do not participate in rapid photochemical reactions below 80 km altitude. The trace gases however undergo a vast number of chemical reactions at all altitudes in the atmosphere. This chemical activity is driven by sunlight, and it is responsible for removal of hydrocarbons, halocarbons, sulfur oxides, nitrogen oxides, and other pollutants which are emitted to the atmosphere by natural or anthropogenic sources. Photolysis of ozone initiates most atmospheric chemistry below 80 km in the atmosphere. The products of ozone photolysis include an electronically excited oxygen atom, $O(^1D)$, which can cleave stable molecules such as H_2O and N_2O to generate free radicals (OH and NO, respectively). These free radicals drive atmospheric photochemistry.

Trace gases also play an important role in regulating the temperature of the atmosphere (climate) since they intercept infrared radiation (heat) at wavelengths where the major gases are transparent. For example, we expect the climate of the earth to warm up if concentrations of nitrous oxide (N_2O) or methane (CH_4) should increase (Lacis et al., 1981).

Nitrous oxide and methane also play important roles in the process that maintains the shield of stratospheric ozone around the earth. Increased levels of N_2O could cause a particularly dramatic decrease of stratospheric ozone (McElroy et al., 1977; National Research Council, 1982, 1984a), with potentially significant ecological impact, since stratospheric ozone prevents damaging ultraviolet sunlight from reaching the earth's surface.

Atmospheric nitrous oxide (N_2O) became the subject of environmental concern during the 1970s, when it was pointed out that modern agriculture may cause a major perturbation to natural processes in the soil (McElroy, 1976; McElroy et al., 1977; Bolin and Arrhenius, 1977) leading to enhanced emissions of N_2O. Combustion of coal (Weiss and Craig, 1976; Pierotti and Rasmussen, 1976) and biomass burning (Crutzen et al., 1979) could also contribute potentially large sources of N_2O to the atmosphere.

Concerns about atmospheric methane (CH_4) also arose in the early 1970s. McConnell et al. (1971) and Levy (1971) pointed out that photooxidation of CH_4 plays a dominant role in the chemistry of atmospheric free radicals and formaldehyde in unpolluted environments. Photooxidation of CH_4 provides an enormous global source of carbon monoxide (CO) and molecular hydrogen (H_2). Examination of likely sources for CH_4 in the atmosphere indicated that rice paddies and domestic cattle provide roughly half of global emissions (Ehhalt, 1974; Ehhalt and Schmidt, 1978), which implies that the atmospheric burden has probably increased substantially in recent history (See Table 5–2). Sze (1978) showed that global atmospheric chemistry is extremely sensitive to changes in the emission rate for CH_4, due to positive feedback between atmospheric concentrations of CH_4 and atmospheric removal mechanisms involving the hydroxyl free radical (OH).

Recent observations confirm the concerns raised about N_2O and CH_4. Global concentrations of both are increasing at the present time (Weiss, 1981; Khalil and Rasmussen, 1983; Craig and Chou, 1983; Blake et al., 1983; Ehhalt et al., 1983) by about 0.2% per year for N_2O and about 1% per year for CH_4 (See Figure 5–1). Atmospheric residence times for both gases are long, about 150 years for N_2O and 10 years for CH_4. A simple analysis of observed trends shows

Table 5–2 CH_4 Budget

Sources	10^{12} gm CH_4/yr	
Enteric fermentation of animals	100–200	agriculture
Paddy fields	280	
Swamps, marshes	190–300	
Fresh water lakes	1–25	
Tundra	0.3–3	
Oceans	1.3–13.6	
Total Biogenic	570–850	
Coal and lignite mining	8–28	
Industrial losses	7–21	
Automobile exhaust	0.5	
Volcanic emissions	0.2	
Total nonbiogenic	16–49	
(from ^{14}C content of $^{14}CH_4$	140–210)	
Total Sources	590–1,060	
Total anthropogenic 400 - 700		

Source: Ehhalt and Schmidt, 1978.

that at present the source for each gas significantly exceeds known removal processes, i.e., the global burden is significantly lower than it would be at steady state. It appears likely, therefore, that the atmospheric concentrations should continue to increase, with important environmental consequences, over the next 10 to 50 years. Indeed data for CH_4 from polar ice cores indicates that concentrations of this gas have doubled since the sixteenth century (Craig and Chou, 1983) representing a massive perturbation to the atmosphere.

Knowledge of the biogeochemistry of these gases is insufficient to assess quantitatively the factors contributing to present atmospheric imbalances, and we have little basis to project the course of future evolution. Only recently have measurements shown that equatorial rain forests are major sources for both gases, using techniques that might allow quantitative estimation of the source strengths (See Figure 5–2). Very little work has been undertaken to establish the data base needed to provide sound estimates of global sources due to agriculture or mining, although a preliminary picture is emerging for rice paddies (Cicerone et al., 1983; Seiler et al., 1984). Virtually all of the ''natural'' sources may change in response to agricultural development or climatic warming.

Carbon monoxide (CO), ozone (O_3), and oxides of nitrogen (NO_x) also play central roles in global atmospheric chemistry. Carbon monoxide and nitrogen oxides are major by-products of combustion. Ozone is produced by interaction of hydrocarbons (or CO) with sunlight, NO_x, and atmospheric radicals and peroxides. These compounds have relatively short chemical lifetimes, from a few weeks for CO and O_3 to a few days for NO_x. Concentrations are consequently

Figure 5–1a. Monthly average concentrations of methane (CH_4) in dry air at Cape Grim, Tasmania (41°N). The Cape Meares data are from continuous measurements, and the Cape Grim data are from large volume samples collected cryogenically. The error bars are one standard deviation of all the measurements for the month.

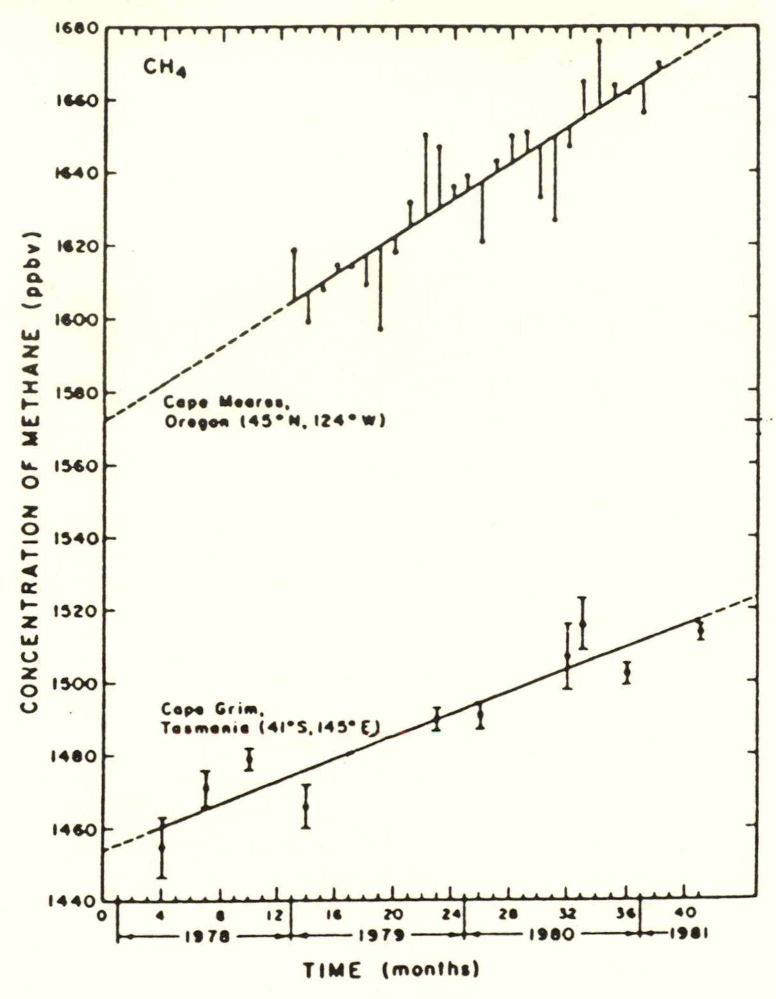

Source: Fraser et al., 1981.

more variable in the atmosphere, and present data provide relatively poor definition of the global distributions and seasonal variations. The data base for CO is generally of high quality but there are major gaps in geographical and seasonal coverage. The data base is extremely sparse for NO_x, while for ozone there are many data of relatively poor quality and with very limited geographical coverage.

Analysis of sources for CO and NO_x indicates a significant role for anthro-

Figure 5–1b. The concentration of nitrous oxide in all 1976–1980 flask samples, normalized to the northerrn hemisphere according to the fit of the data to the combustion source model, including the latitudinal term with ϕ_0 = 5°N, and plotted against time together with the fitted curve for the northern hemisphere.

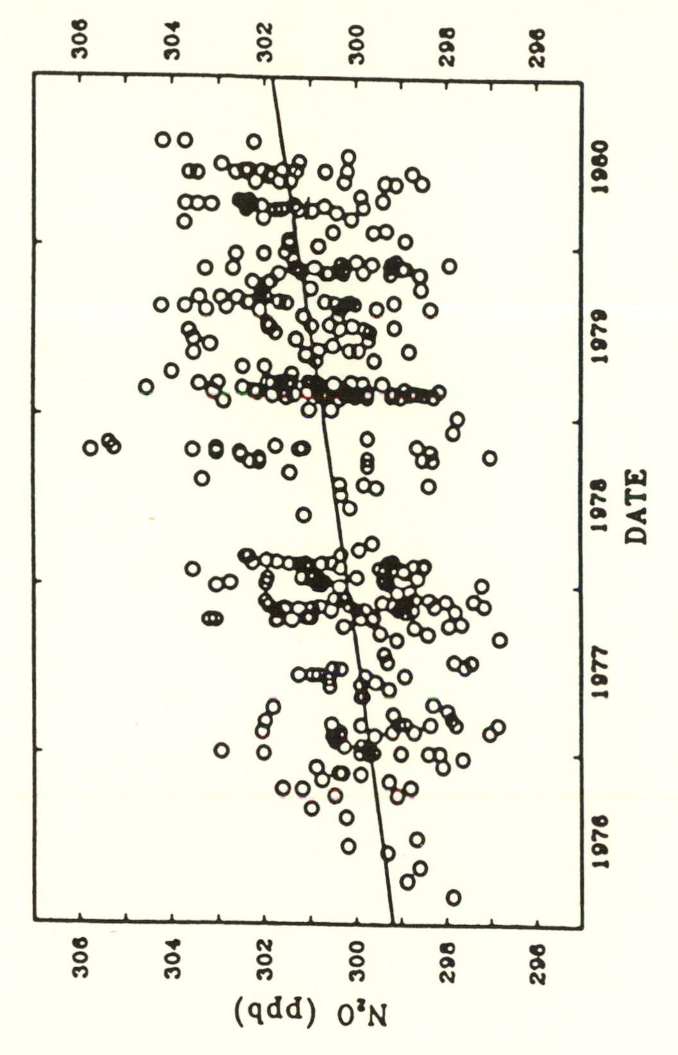

Source: Weiss, 1981.

Figure 5–2a. N_2O concentrations in parts per billion over Brazil, showing enhanced concentrations over central South America. Analysis of these contours using a general circulation model for winds indicates a globally significant source for N_20 from the rainforests of Brazil.

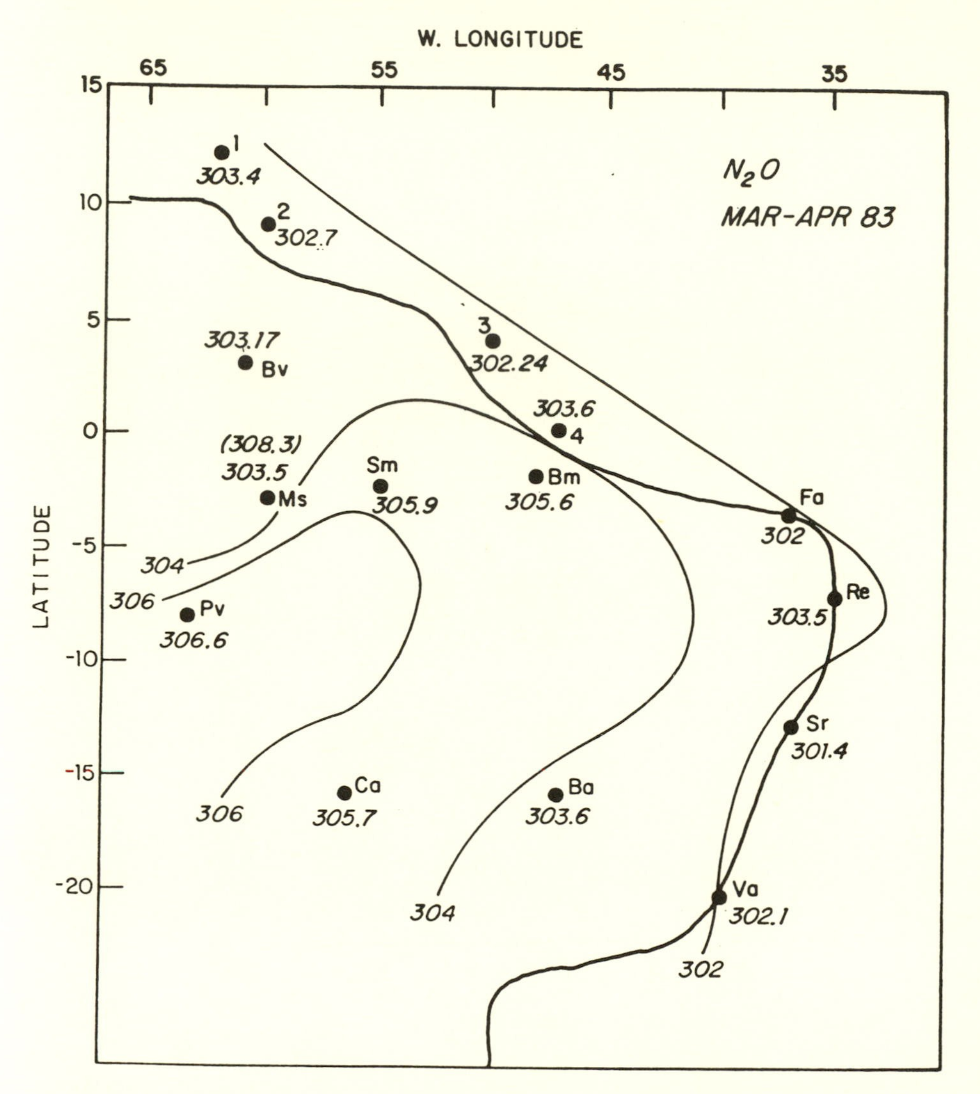

pogenic emissions, mainly combustion (cf. Logan et al., 1981; Logan, 1983; Khalil and Rasmussen, 1984). We therefore expect that present concentrations of these species may be significantly elevated over preindustrial values. The magnitudes of such increases remain uncertain. Carbon monoxide is a by-product

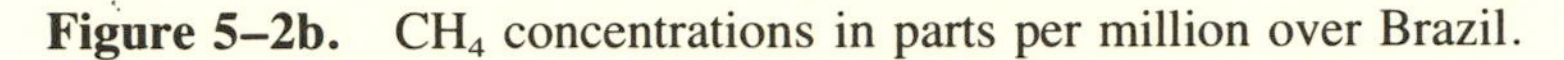

Figure 5–2b. CH_4 concentrations in parts per million over Brazil.

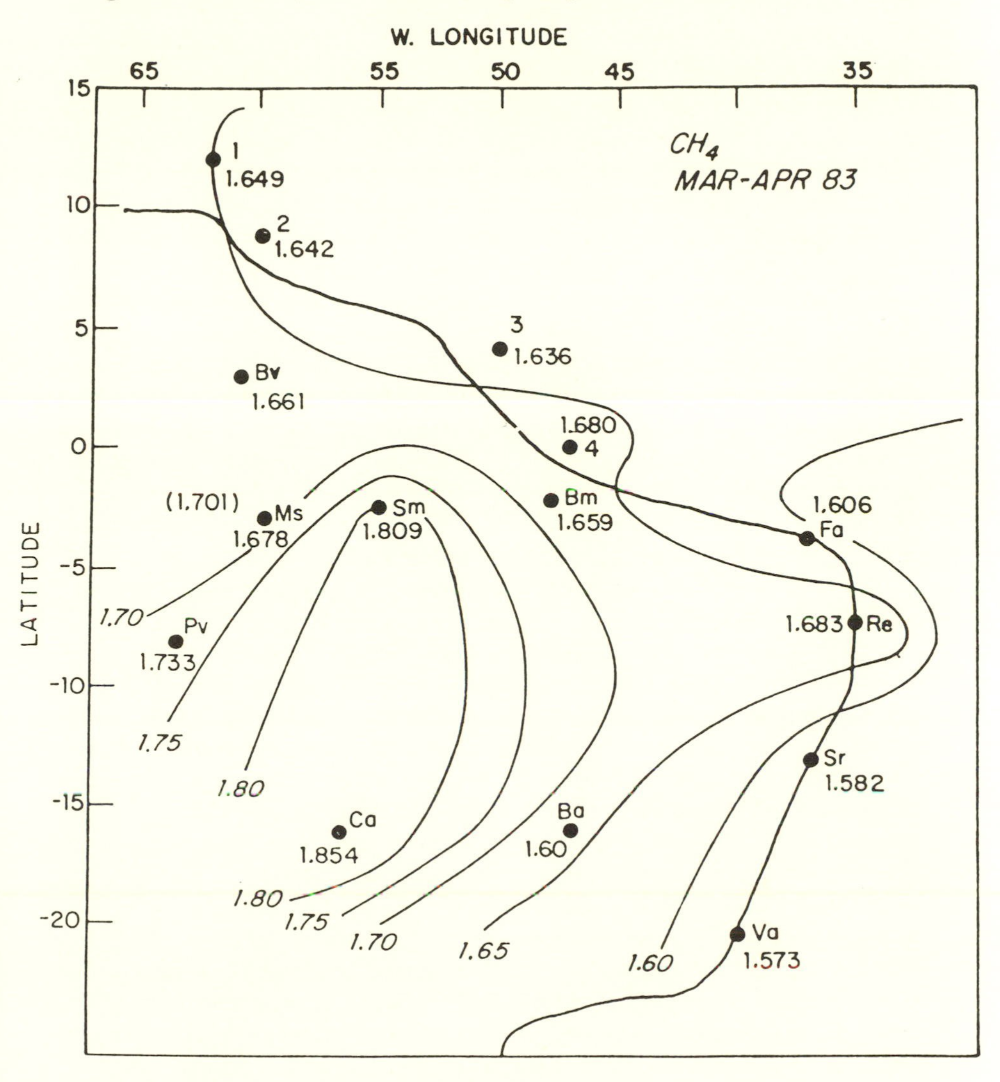

Source: Wofsy, unpublished manuscript, 1984.

of photooxidation of natural hydrocarbons and CH_4, but the importance of this source is highly uncertain. Oxides of nitrogen are produced by soils (Galbally and Ray, 1978; Lipschultz et al., 1981), but very little is known about the possible significance of this process. Lightning may be a major natural source of NO_x. Recent estimates for the magnitude of this source range over more than an order of magnitude (cf. review by Logan, 1983).

Stratospheric Chemistry

Stratospheric ozone provides a screen which prevents biologically damaging ultraviolet radiation from reaching the surface of the earth. Ozone is maintained by a complex suite of photochemical reactions, and its concentration is potentially very sensitive to homogeneous catalysis by free radicals such as NO, ClO, or OH. Concern over possible damage to the ozone layer arose during the 1970s, initially focused on effluent from commercial stratospheric aircraft, and later focused on industrial halocarbons and on perturbations to atmospheric N_2O. NASA and NSF have supported vigorous research programs on stratospheric chemistry and substantial progress has been made, as may be seen from the progression of NRC reports on the subject (National Research Council, 1973, 1975, 1976a, b, 1977, 1979, 1982, 1984a). Efforts to measure concentrations of key species, to determine important reaction rates, and to model complex dynamics and photochemistry have been particularly impressive. Stratospheric research programs stimulated kineticists to master the difficult art of measuring radical-radical reaction rates. Kinetic studies have revealed previously unsuspected behavior at pressures appropriate to the atmosphere. Important free radical species have been measured in the atmosphere, a feat widely considered impossible in the early 1970s.

Stratospheric chemistry is driven by inputs of solar ultraviolet radiation and by long-lived, insoluble trace compounds transported upward from the troposphere. Short-lived tropospheric species and soluble gases are removed before reaching the stratosphere. Once injected into the stratosphere, however, a given atom or molecule may reside there for a long time, up to several years. The major sources of HO_x and NO_x radicals in the stratosphere are reactions of excited oxygen atoms $O(^1D)$ with H_2O and N_2O, respectively. Water is derived in part from upward transport through the tropopause, but oxidation of CH_4 also provides a significant contribution. The major source of Cl and ClO radicals is photodecomposition of long-lived anthropogenic chlorocarbons (Freons, CFC–11 ($CFCl_3$), CFC–12 (CF_2Cl_2) and 1,1,1,-trichloroethane (CH_3CCl_3). Methane also plays a key role as a reactant with OH and Cl radicals. Thus the chemistry of the stratosphere depends on long-lived gases which originate in the troposphere and which are known or suspected to be strongly influenced by human activities.

There are major gaps in our understanding of stratospheric chemistry and dynamics, and hence there are significant uncertainties in estimates of the ozone response to given changes in its atmospheric composition. Our knowledge of long-term trends in stratospheric composition is also inadequate to define the extent to which the stratosphere is presently changing. Ozone itself provides the best example. We have at present a scattered, somewhat sporadic network of ozonesonde stations which use an instrument with many inherent problems, and there are satellite data with global coverage for limited time periods and with coarse resolution. The data show that O_3 concentrations vary markedly in space and in time, and hence global spaceborne observations are required to define the distribution accurately. It is difficult and expensive however to maintain the required absolute precision in a spaceborne instrument over a long period of time. NASA is attempting to do this by using space shuttle flights to recalibrate instruments on the upcoming UARS satellite. However, natural perturbations such as the eruption of E1 Chichon volcano make it very hard to detect pollution-induced trends, even with a perfect instrument. A long-term commitment to surveillance of the stratosphere is required—an expensive and arduous task.

Tropospheric Chemistry—Continental and Rural

Considerable research effort has been devoted to studies of tropospheric pollution on the urban scale. Much less attention has been given to studies of the chemistry and transport of pollutants over intermediate scales, for example the continental or regional scale, although measurements have shown that secondary products of urban air pollution (ozone, nitric acid, and sulfate) have significant influence on air and water quality on these scales.

The atmosphere contains a variety of minor stable hydrocarbons such as ethane (C_2H_6), actylene (C_2H_2), propene (C_3H_6), and propane (C_3H_8) (Rudolf and Ehhalt, 1981). Concentrations are large enough to be important to atmospheric chemistry in clean environments, and significant enhancements are observed in areas affected by long-range transport of air pollution. The atmosphere over populated continents probably also contains important concentrations of labile species such as aldehydes, ketones, alcohols, organic acids, peroxides, and organonitrogen compounds, but the present data base consists of only a few measurements of selected species from each group. These are in general reactive, phytotoxic chemicals, produced in abundance during photodecomposition of anthropogenic hydrocarbons (Atkinson et al., 1982). As yet we know very little about the concentrations, chemistry, and impacts of such species in the environment. Preliminary information suggests potentially significant effects.

Deposition patterns for nitrate ion (NO_3) show large regional excesses, as may be seen in Figure (5–3b). The nitrate data imply corresponding enhancements for reactive NO_x species in the atmosphere over the continent (Fishman, 1981; Logan, 1983). It is important to note that the deposition pattern for sulfate ($SO^{=}_4$) is very similar to that of NO_3 (See Figure 5–3a), indicating a close association of these two acid-forming species and hence implying a close association between acid deposition and elevated concentrations of nitrogen oxides and other phytotoxic chemicals. Acid deposition is widely blamed for forest decline in Europe and the eastern U.S., but as yet the specific mechanisms responsible have not been elucidated. It can be reasonably concluded that toxic effects of acidity should be concentrated geographically in the same areas as impacts of NO_x, O_3, and other labile chemicals, and it is therefore difficult to distinguish the effects of acidity from those of phytotoxic chemicals (Dunnet, 1983).

Figure 5–3a. Spatial distribution of mean annual wet deposition of sulfate weighted by the amount of precipitation in North America in 1980 (mmoles/m^2)

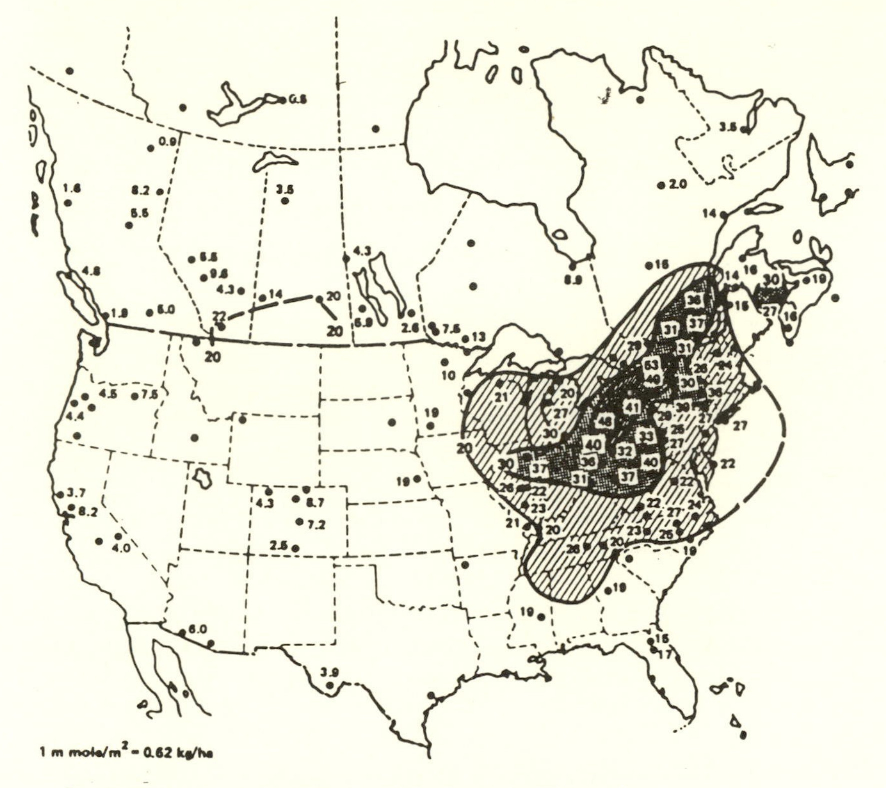

Source: U.S./Canada Work Group #2 (1982); from NRC, 1983.

The possible importance of reactive pollutants in rural and forest areas becomes particularly clear when we examine ozone distributions over the continents. Monthly mean ozone concentrations from Australia and Canada usually fall in the range of 15–30 ppb near the ground. Rural sites in North America and Europe, however, show a pronounced summer maximum with much higher ozone levels, between 35 and 50 ppb (Figure 5–4). Data from the GDR indicate increases in summer mean concentrations from 17 ppb in 1956 to 35 ppb in 1976 at a rural site on the Baltic Sea (Warmbt, 1979) (See Figure 5–5) and data from Scandinavia indicate major ozone episodes originating on the European continent (Hov et al., 1978; Isaksen et al., 1978). Ozone episodes regularly blanket the entire eastern U.S. with concentrations in excess of 100 ppb (See Figure 5–6; Wolff et al., 1977a, b; Wolff and Lioy, 1980; Vaughan et al., 1982; Samson and Ragland, 1977). These data imply an approximate doubling of summer O_3 concentrations since the 1950s and indicate episodic levels high

Figure 5–3b. Spatial distribution of mean annual wet deposition of nitrate weighted by the amount of precipitation in North America in 1980 (mmoles/m^2)

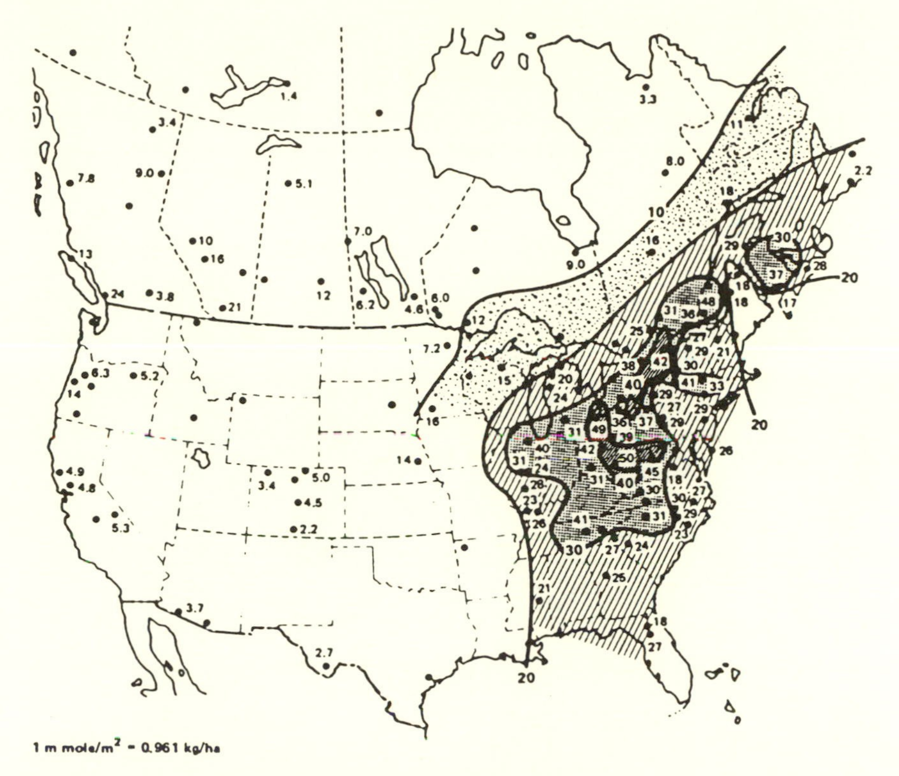

Source: U.S./Canada Work Group #2, 1982; from NRC, 1983.

enough to severely injure plants (Heck et al., 1982; Dunnett, 1983; see Figure 5–7). There are clearly potentially major economic impacts on agriculture, and possible impacts on forests are unknown.

The atmospheric chemistry of acids is closely linked to atmospheric oxidants and hydrocarbons. The major oxidant nitrogen dioxide (NO_2) is oxidized by the OH radical to nitric acid, one of the major sources of acidity in precipitation and dry deposition (NRC, 1983). The other major acid, H_2SO_4, is produced by gas-phase oxidation of SO_2, involving OH, and by aqueous phase oxidation, involving H_2O_2 (NRC, 1983). Abundances of both OH and H_2O_2 are controlled by the suite of pollutants present in air over populous continental areas. Concentrations of O_3 in remote areas is quite sensitive to the ambient levels of NO and NO_2 (Crutzen, 1979; Logan et al., 1981).

Efforts to study either the chemistry or the effects of atmospheric acids must inevitably be broadened to include reactive gaseous pollutants. The linkage between these two types of pollution is so close that, from a conceptual stand-

Figure 5–4. (Monthly Mean) Ozone data at the ground for rural sites in U.S. and Europe are compared to rural sites in Australia and Canada. Southern hemisphere data have been shifted 6 months. Note contrasting seasonal patterns and elevated concentrations in Europe and U.S.

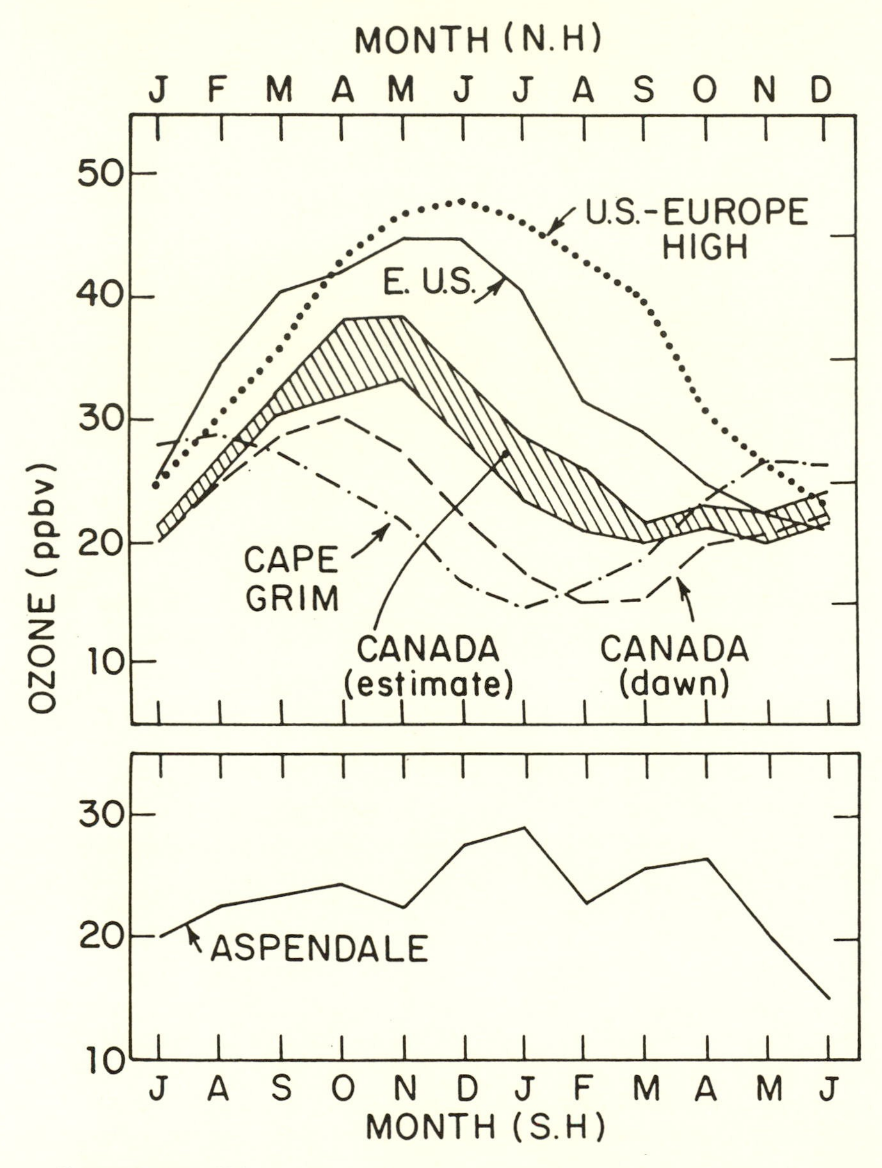

Source: Logan, 1984.

Figure 5–5. Ozone Concentrations at rural site in the GDR, on the Baltic Sea. Note the doubling of summer values from 1956 to 1978, so that present values are similar to those of rural sites in the USA and other parts of Europe.

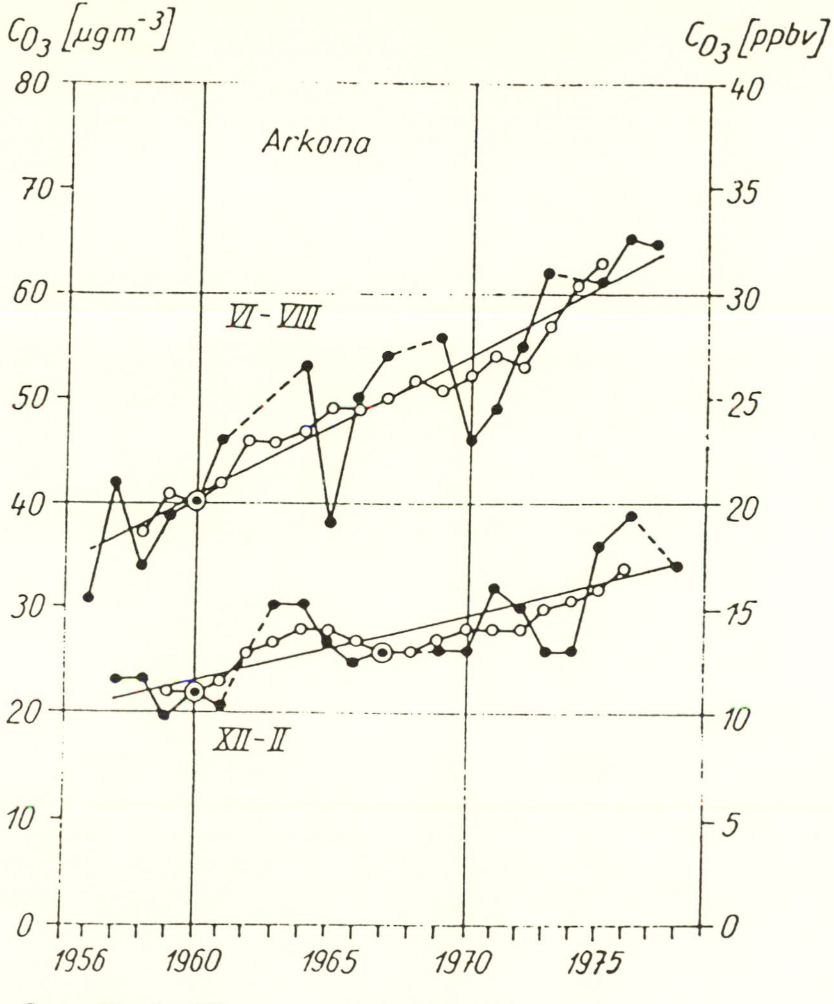

Source: Warmbt, 1979.

point, they must be regarded as manifestations of a single phenomenon, continental-scale air pollution. If we are to understand and to mitigate the problems associated with pollution on this scale, an integrated conceptual framework is required which addresses acids, oxidants, radicals, hydrocarbons, and other important species. Present research programs are narrowly focused on acid deposition, probably because acidity is rather easy to measure while phytotoxic

Figure 5–6. Air parcel trajectories during July 14–20, 1977; (b-d) ozone concentrations patterns

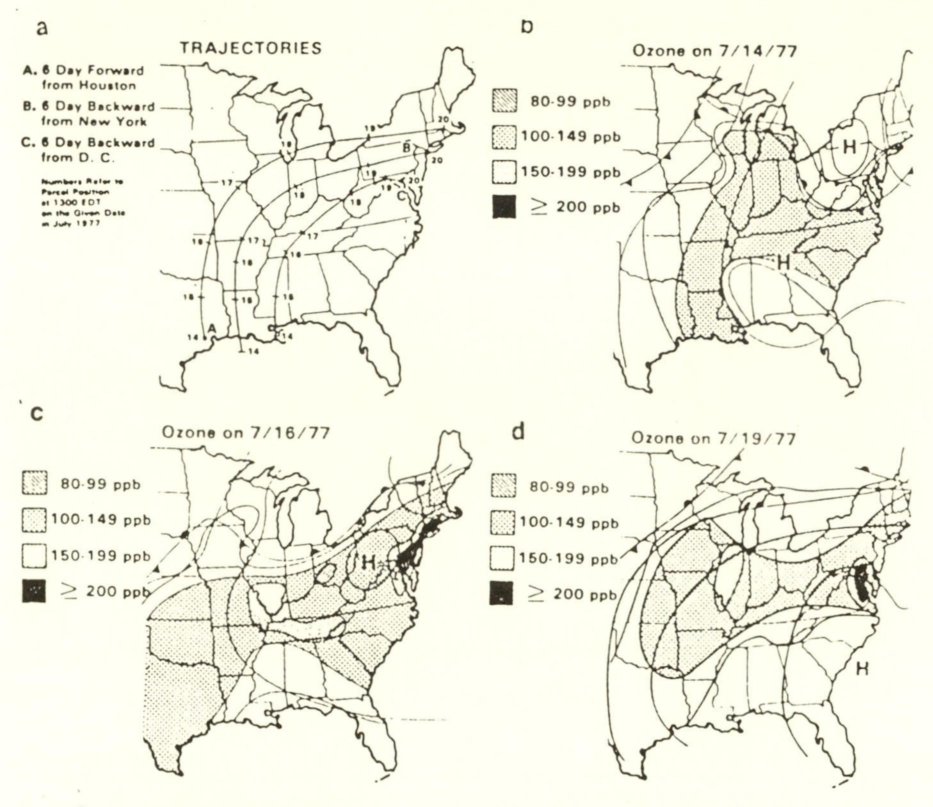

Source: Wolff and Lioy, 1980

gases are very difficult to measure. It is unlikely, however, that such a narrow approach can provide the necessary understanding of factors which contribute to damage of crops and forests. The need for a broadbased study was in fact recognized some time ago, as may be seen by examining the report.

LONG-TERM ENVIRONMENTAL RESEARCH—OPPORTUNITIES

If we consider time scales of five years or longer, the atmosphere, biosphere, and oceans function as an integrated system. Physical, chemical, and biological processes are coupled, and progress in understanding them requires interdisciplinary research efforts of broad scope and content. These essential facts should provide the conceptual framework for programs of research, in order to assure optimum return on invested resources.

The discussion of science background above focuses attention on two spatial

Figure 5–7. Dose-response curves for yield of five plant species as a function of the seasonal 7-h/day mean O_3 concentration. Yield is given as the percent of the yield predicted at 0.025 ppm O_3 by a simple linear model (●) and a plateau-linear model where applicable (O). (O) indicates a coincidence of data points. The linear model is shown as a solid line and the plateau linear as a dashed line.

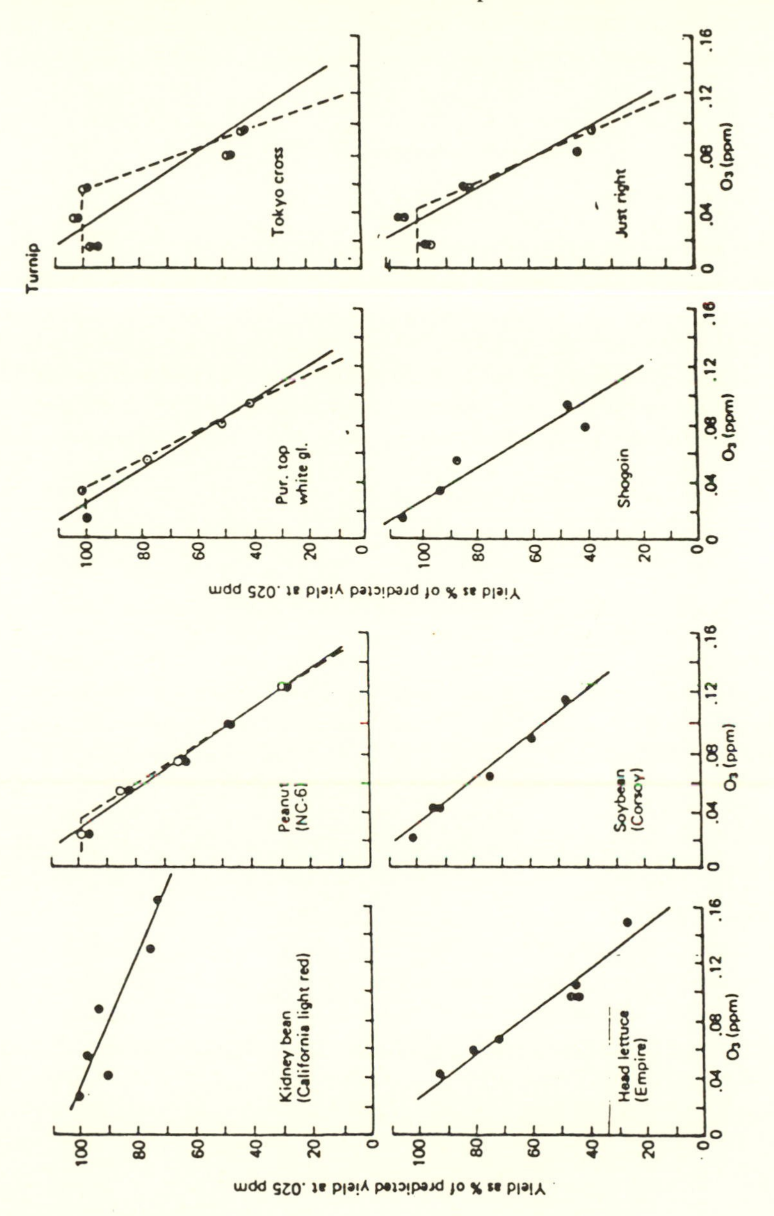

Source: Heck et al., 1982

(and temporal) scales: global problems and continental scale problems. Changes in the global atmosphere raise novel issues of profound ecological and environmental significance. New, creative approaches to research are required to address these issues, which cut across many disciplinary and administrative boundaries. High concentrations of pollutants on the continental scale, in rural and forest areas, raise the threat of direct economic and health effects as well as more general environmental concerns.

Global Atmospheric Chemistry

The framework for a major research initiative has been recently elaborated in *Global Change: A Biogeochemical Perspective* (McElroy et al., 1983). The initiative, called the "Global Habitability Program," is intended to encompass a variety of ongoing research programs in NASA, NSF, EPA, NOAA, and DOE. In the view of the author, this program addresses the fundamental need to attack long-term large scale environmental problems with a coordinated research effort with global perspective. The program brings together and strengthens presently disparate efforts. The following is a summary of the coordinated program discussed in *Global Change: A Biogeochemical Perspective*.

Program Definition

The program focuses on major biogeochemical cycles (C, N, P, S, H_2O), and on the range of factors judged important for the global life support system (solar radiation, the quality of air, soil, and water, and climate). Man's influence on biogeochemical cycles is apparent and widely recognized. Perturbations to the carbon cycle are manifest through readily observable changes in the concentrations of atmospheric CO_2 and CH_4. Likewise, perturbations to the nitrogen cycle are evidenced by changes in concentrations of atmospheric N_2O and suspected changes for other oxides of nitrogen. Increased mobilization of sulfur leads to enhanced concentrations of sulfate in precipitation and to an additional burden of sulfate aerosol in continental haze and arctic haze. Combustion of coal and operation of diesel engines produce carbonaceous fine particulates with long atmospheric residence times and with high radiative and biological activity.

Tropospheric Research. There are five major goals for tropospheric research with a global perspective: to understand the principal components of the hydrological cycle; to define the processes which regulate the distribution and abundance of atmospheric oxidants, notably O_3, SO_2, and NO_2; to define factors that influence the chemistry of acid deposition, both dry and wet; to identify processes which affect the abundance of aerosols and of gases such as CO_2, N_2O, CH_4, NH_3, H_2O; and to assess the impact of anthropogenic infrared absorbers on the radioactive budget of the atmosphere and on climate.

These goals may be translated into a number of specific research objectives.

We need observational data and associated laboratory and theoretical analyses to define the extent of the human influence on the global distribution of tropospheric O_3 and other oxidants, aldehydes and other phytotoxic chemicals, and mineral and organic acids. Research on long-lived biogenic gases (CH_4, N_2O) is needed to understand the factors which control global biological and industrial emissions and to define quantitatively mechanisms for chemical degradation in the atmosphere.

The radical OH plays a central role in the removal of tropospheric gases such as SO_2, H_2S, CH_3Cl, CH_3CCl_3, CO, CH_4, and other hydrocarbons. We see an urgent need for instrument development, and for field, laboratory, and theoretical studies to define the chemical processes which determine the abundance and distribution of OH in the troposphere. Information on OH, in combination with measurements defining the distribution of short-lived C- , N- , and S-bearing gases (CO, $(CH_3)_2S$, etc.) can provide valuable clues to the function of the major biogeochemical cycles, contributing significantly to the goals stated earlier for the troposphere.

Stratospheric Research. The primary goal for stratospheric research is to understand the processes that regulate the distribution and abundance of O_3. The focus is on the impact of man, not only as it may affect transmission of ultraviolet radiation to the earth's surface, but also as it may influence regional and global climate.

The goals for stratospheric and tropospheric research are not unrelated. Ozone and the radicals OH and NO play central roles in both regions of the atmosphere. Chemical processes in the troposphere significantly influence the supply of gases to the stratosphere. Methylchloride, for example, is thought to provide the largest source of chlorine in the unperturbed stratosphere. The quantity of CH_3Cl reaching the stratosphere is regulated by reactions involving OH in the troposphere and by biogenic and anthropogenic emissions. Similar processes regulate the flux of CH_4 to the stratosphere. The stratosphere provides an important source of tropospheric O_3. Ozone is both produced and consumed in the troposphere, with sources and sinks modulated, it is thought, by the ambient level of NO_x.

Atmosphere-Ocean Interactions. The open ocean, coastal waters, and coastal wetlands can provide important sources for atmospheric gases such as CH_4, H_2S, and $(CH_3)_2S$ and the coastal environment is thought also to play a significant role in production of halocarbons such as CH_3Cl. Research goals for coastal systems should focus on improving our understanding of processes as they affect coastal marine productivity and as they influence transfer of important chemical species from ocean to atmosphere.

The general research goal for the oceans is to quantify and understand the role of the oceans in climate and in atmospheric chemistry. It is particularly important to understand the mechanisms that regulate primary production, nitrogen fixation, denitrification, and the burial of organic debris in the sea. The coupling of physical and chemical processes across the atmosphere-ocean boundary, and across the thermocline, influences the availability of nutrients and affects the rate of photosynthesis and rates for biogenic production of gases (N_2O, $(CH_3)_2S$, CO_2, etc.) in the upper mixed layer.

Interactions between the Atmosphere and the Terrestrial Biosphere (including soils). We need to specify accurately the extent and elemental content of major terrestrial ecosystems (tropical forests, savanna, cropland, etc.) and to monitor these systems for change over an extended period. Studies of perturbed systems, regions of major land clearance for example, may be particularly instructive in this regard. The major questions that need to be addressed are: To what extent is the storage capacity for carbon nutrients and metal ions influenced by changes in atmospheric inputs of acids and nutrients, by changing levels of atmospheric oxidant, or by changes in local climate? How do changes in the biota feed back to modify the chemistry of air and water, the hydrologic cycle, or climate? What is the role of biomass burning in global atmospheric chemistry? We see the need for an extensive program of remote sensing, in combination with intensive studies of selected systems, to address these issues. Observational programs must be complemented by appropriate theoretical field, and laboratory investigations. Studies of the biota are by their nature complex, requiring an interdisciplinary approach, with cooperation between physical and biological scientists essential if we are to make progress.

Investigation of a perturbed tropical ecosystem provides a case in point. We see the need for in situ atmospheric and aquatic measurements to define the input and removal of important chemical species from a selected study area (e.g., Amazon Basin) with an appropriate control. We see a requirement for in situ investigations to identify the nature of relevant internal biological processes, with remote sensing employed in order to place the local investigation in context, permitting extrapolation to larger scale. Investigation of tropical systems could be imbedded thus in a larger matrix of studies designed to improve our understanding of the overall metabolism of the biosphere.

Long-Term Environmental Research Elements: Particular Problem Areas Related to Global Issues

Long-term baseline measurements. Several types of long-term environmental research can be identified to aid development of the Habitability Concept. Some of these are outlined in a recent report of the Board of Atmospheric Science and Climate (National Research Council, 1984b). One of the most important, and most problematical, is long-term baseline studies of atmospheric gases. These studies are needed to define the course of chemical change in the atmosphere and to elucidate seasonal changes and synoptic-scale fluctuations. The data also provide unique information about the factors that influence gas concentrations, and they are essential for rigorous testing of realistic, three-dimensional models of the atmosphere. If baseline data are lacking or of poor quality, even a very significant perturbation to the atmosphere may evade detection. The studies cited above illustrate the painstaking work required to reconstruct even relatively large changes in atmospheric composition, for example, increases in tropospheric methane or ozone concentrations.

Despite the importance of long-term atmospheric measurements, present programs are beset by major problems. The existing network for ozone is seriously inadequate, both in terms of data quality and geographical and temporal coverage. There is no long-term data base for NO_x at all, due in part to analytical difficulties, which may have been resolved recently. Some important gases, such as N_2O, $CFCl_3$, and CH_3CCl_3 have been studied in the recent Atmospheric Lifetime Experiment (ALE), supported by NASA and by the Chemical Manufacturers Association (Prinn et al., 1983). This network provided excellent data, with a measurement every four to eight hours for five years at four stations. The results show dramatically that high frequency sampling is, in fact, needed to separate atmospheric variations from analytical fluctuations, even for the longest-lived gases. Unfortunately, the ALE stations might not continue in operation. Baseline studies (''monitoring'') are considered by some to be inappropriate activities for support by research agencies, despite unmistakable historical evidence showing that successful baseline studies are scientifically challenging problems best carried out by individuals with a research orientation. Indeed the ALE-type program needs to be extended to a larger number of stations and to other gases, and hence a significant commitment of funds and scientific talent is required.

NOAA does have a monitoring program, using an excellent array of sites, called Global Monitoring for Climatic Change. Unfortunately, despite the enormous potential of this program, good data have not been forthcoming and will likely not be produced without major reforms. A program called RITS (Radiatively Important Trace Species) has been proposed by NOAA's Environmental Research Laboratory to provide high-quality ozone data and to begin the reform process for GMCC. If this program does not proceed and if NASA cannot fill the gap, we can anticipate continued failure to meet the scientific needs for global monitoring data.

The need for high-frequency, accurate baseline data was identified some time ago by atmospheric scientists. The last decade represents a number of lost opportunites in this regard and there does not seem to be any assurance that the next decade will be better.

Biogeochemical research. Studies of source processes for important gases in major ecosystems are clearly essential to obtain an understanding of atmospheric chemistry. Here we are thinking about ground-based and aircraft experiments designed to measure fluxes of important species from key environments, including tropical forests, savannas, agricultural areas, tundra, areas with large-scale biomass burning, and even urbanized regions of continents (see National Research Council, 1984b). A number of such experiments have been undertaken in the USA by NASA (for example, Global Tropospheric Experiment), NSF (through grants and through the National Center for Atmospheric Research),

EPA (urban plume studies), National Acid Deposition Assessment Program (NADAP), and Electric Power Research Institute (mainly sulfur emission studies). A number of scientists in Europe have also undertaken such studies.

It is clear that this type of research is at an early, exploratory stage. The committee examining the Habitability concept perceived the need for a more systematic approach, with emphasis on coordinated, interdisciplinary investigations and with expanded levels of support. There have been some indications of movement in this direction recently and these should, in the opinion of the author, be encouraged and stimulated. Some novel and important questions have been posed about gas fluxes from (or to) major ecosystems and from anthropogenic activities. Investment in such research is likely to provide high return.

Laboratory investigations and instrument development. Laboratory studies provide major stimulus for new concepts. Studies of chemical kinetics at atmospheric pressure have revealed previously unsuspected influence of pressure on key reaction rates, and important new transient compounds have been discovered. Fundamental research on microbial metabolism has significantly aided interpretation of gas emissions observed in natural systems, and new analytical methods for sensitive atmospheric measurements have repeatedly stimulated important advances in the understanding of atmospheric chemistry. Programs to support this sort of work have been in place at NASA, NOAA (ERL), and NSF, and support has also been provided by private sector organizations such as CMA. This work seems to be on a relatively sound footing, hence few additional comments are required here. We note that instrument development is a capital intensive activity that would benefit from input of additional resources. Laboratory studies of heterogeneous chemistry and liquid-phase chemistry in droplets are urgently needed. These are difficult to perform, but may be very important for understanding certain processes such as oxidation of SO_2 to H_2SO_4.

Continental-scale Air Pollution

The conceptual framework for studies of continental-scale pollution is in many ways similar to that discussed above for global scale atmospheric chemistry. Many of the key species are the same (O_3, CO, NO_x, OH). It should not be assumed that Europe and North America are the only regions where these problems are important. Arctic haze affects nearly the whole polar region in winter (Schnell and Raatz, 1984; Raatz and Schnell, 1984) and agricultural burning impacts vast areas of Savanna in South America and Africa (Crutzen et al., 1984). China, India, and other populous areas are probably affected to some degree by large-scale pollution.

There are, of course, major differences between global-scale chemistry and rural chemistry in populated continents. Trace pollutant concentrations are 10 to 100 times higher in the latter case, although these levels are still much lower than those encountered in urban areas. A variety of relatively short-lived (1 day–1 week) compounds are more important in continental-scale pollution.

Program Definition—Continental Scale

The program should focus on four goals:

1. to define and to understand the factors controlling fluxes of important gases and aerosols from (or to) urban source regions, agricultural areas, and undisturbed areas;
2. to determine chemical concentrations, transformation rates, and transport rates for pollutants on scales of 100–10000 km;
3. to elucidate mechanisms and rates for deposition of pollutants at the earth's surface, and to define the fate of the deposited material; and
4. to investigate short- and long-term effects of acid deposition and toxic chemicals on plants, animals, soils, and waters in populous continental areas.

This conceptual framework is analogous in many ways to that outlined above for global atmospheric chemistry, (McElroy et al., 1983; National Research Council, 1984b) and to plans discussed in connection with the National Acid Deposition Assessment Program (NADAP cf. National Research Council, 1983) for North American acidic pollution. The main distinction is the integrated, coordinated approach that we advocate to include major pollutants on relatively large spatial scales. The rationale for a program that encompasses the suite of major pollutants has been stated clearly above. This broad based approach is, in the opinion of the author, essential for success in understanding and ameliorating the large-scale environmental impacts of air pollution. It should be noted that worldwide increases in population and in the level of industrialization promise to inexorably expand the regions of the earth affected by continental-scale air pollution. As this trend continues we may anticipate gradual blurring of the distinction we have made between atmospheric chemistry on global and continental scales. The similarity of conceptual framework reflects the underlying convergence of these problems.

Research Elements—Problem Areas Related to Continental-Scale Air Pollution

Baseline Data. Good baseline data are as important for continental pollution studies as for research on global atmospheric chemistry. However, there are almost no programs directed toward obtaining such data, and no agency accepts the responsibility for supporting or undertaking such measurements. Reliable long-term records are lacking even for the simplest parameter, the pH of rain. Nevertheless the recent NAS report on acidic deposition (National Research Council, 1983) does not even mention baseline studies in its consideration of research needs. At present, an effort to measure pH appears to be under way at last. The GDR ozone network cited above (Warmbt, 1979) represents one of the few measurement programs designed to determine long-term trends for trace gas concentrations in nonurban continental air. Rural ozone measurements in the U.S. have for the most part been run for a few years, then discontinued (cf. Evans et al., 1983). We are left therefore without observations to define present trends or to determine the current impact of pollution on ozone levels.

Other long-term data have been obtained by serendipity. Graedel and McRae (1982) used unpublished monitoring data from New Jersey to infer the long-term trend in atmospheric CH_4. It is quite surprising that the quality of the data was adequate for this purpose. The ALE station at Adrigole, Ireland, was intended to give a record of Freon concentrations in clear air from the North Atlantic. Instead it provided a unique, high-resolution record of frequent episodes of pollution advected over hundreds of kilometers. This station has now been closed down, but the data base is providing a rich source of information for researchers interested in verifying three-dimensional models or in studying the morphology of pollution events.

Baseline measurements provide an essential part of any program to study continental-scale air pollution. These data are needed to define the current state of the atmosphere, to shed light on important source mechanisms, and to allow determination of long-term trends. Some surprises are likely. For example, unpublished data for 1973–82 by Mohnen (private communication, 1983) at Whiteface Mountain, New York, show interesting seasonal cycles in ozone concentration and in the variance of ozone concentration: both maximize in summer, when monthly mean ozone exceeds 50 ppb. Any model purporting to simulate large-scale pollution chemistry should be able to reproduce the major features of ozone climatology, represented here unfortunately by very few stations. The Whiteface Mountain data show a maximum in 1978; was the subsequent decline due to the beneficial effects of emission controls on automobiles? Given the resources expended to combat air pollution, we should be able to examine such crucial questions with an array of long-term sampling data. Such data do not exist and are not being collected at present.

Instruments at baseline stations must be maintained for an extended period ($\sim$10 years) with stable calibration. Measurements should include as many pollutants as possible, such as O_3, NO, NO_2, SO_2, and some hydrocarbons. Measurements should be made as often as possible in order to average down instrumental errors and to resolve diurnal and synoptic scale variations. The large number of data points require use of automated instrumentation and computerization of results. The ideal number and location of sites cannot be specified until pilot stations have been operated for some time, to define the important spatial and temporal scales. We can say however that sites should be 100 km or more distant from urban centers or major industrial complexes, in order to avoid enormous fluctuations due to nearby sources.

Urban Plume Studies. It is important to define the emission rates for moderately long-lived pollutants such as NO_x, aldehydes, or PAN. This is a difficult task, because primary emissions are concentrated in urban areas where photochemical activity is intense. The emitted gases may be substantially modified while being transported away from the urban source region. For example, NO_x may be lost and unsaturated hydrocarbons may give rise to significant concentrations of aldehydes, ketones, or PAN. Thus urban plume studies are analogous to the biogeochemical source studies discussed in the section on global atmospheric chemistry.

A number of field studies have been undertaken by EPA to examine the chemistry of an aging urban plume (Spicer, 1977; Spicer et al., 1979, 1982 a, b; Spicer, 1982; Possiel et al., 1982). These experiments have provided useful data, although the focus has generally been on the chemistry in the plume, rather than on the flux of pollutants into the larger-scale circulation. It would be very useful to have additional urban plume studies using more sophisticated instrumentation, to look at a wider range of pollutants. Innovative experimental design is needed to allow examination of heterogeneous processes. It would be helpful also to shorten the time delay between the execution of these studies and publication of results.

In Situ Studies of Labile Species. Labile molecules and radicals play crucial roles in atmospheric photochemistry. Field-suitable measurement techniques for species such as OH, HO_2, PAN, alcohols, and peroxides are needed in order to understand the transformation processes in the atmosphere and to define the phytotoxicity of the pollutant mix. Efforts to develop good field instruments are ongoing, and the need is mentioned here mainly to emphasize their importance.

Theoretical Studies. Models of the atmosphere have traditionally provided the conceptual framework for pollution studies, and have provided guidance for much of the experimental work. Existing models generally use a hybrid approach between parameterized empirical treatments and ab initio theoretical attack on the problem. The long-range component of theoretical work lies mainly in the development of realistic, three-dimensional models with proper physical representations of both transport and chemistry. There are four applications for such advanced models: reconstruction of significant events, simulation of pollution climatology, prediction of future atmospheric conditions, and assessment of various control or development strategies. The event study is intended to elucidate factors that contribute to development of major pollution episodes, e.g., to determine where the most important ozone precursors entered the system, or to analyze how a given event would have developed if various options for emission reduction had been adopted. The climatology models are intended to elucidate the factors that control mean and maximum concentrations or deposition rates at rural sites, to examine the effect of emission control scenarios on potential crop damage, etc.

Mechanistic models of pollution phenomena are invaluable because such models can predict the important characteristics of the system under altered conditions for primary emissions, different distributions of industrial activity, etc. Parameterized models cannot, in general, perform such an assessment reliably. Mechanistic models also allow rigorous testing of proposed kinetic reaction schemes. It should be recognized, however, that atmospheric dynamics are multiscaled and nonlinear and are not understood completely. Hence development of realistic ab initio models must be regarded as part of a long-term research program, and not treated as a straightforward application of known principles. In the author's opinion, major scientific returns will be realized by the ongoing efforts to create and improve such models.

REFERENCES

Atkinson, R., A. C. Lloyd, and L. Winges, 1982. An updated chemical mechanism for hydrocarbon/NO_x/SO_2 photooxidation suitable for inclusion in atmospheric simulation models. Atmos. Environ. *16,* 1,341–355.

Blake, D. R., E. W. Mayer, S. C. Tyler, Y. Makide, D. C. Montague, and F. S. Rowland, 1983. Global increase in atmospheric methane concentrations between 1978 and 1980. Geophysical Res. Letters *9,* 477–80.

Bolin, B. and Arrhenius, 1977. Nitrogen: an essential life factor and growing environmental hazard. Ambro *6,* 96–105.

Cicerone, R. J., J. D. Shetter, and C. C. Delwiche, 1983. Seasonal variation of methane flux from a California rice paddy. J. Geophys. Res. *88,* 11,022–024.

Craig, H. and C. C. Chou, 1983. Methane: the record in polar ice cores. Geophys. Res. Lett. *9,* 1,221–224.

Crutzen, P. J., L. E. Heidt, J. P. Krasnec, W. H. Pollack, and W. Seiler, 1979. Biomass burning as a source of atmospheric gases CO, H_2, N_2O, NO, CH_3Cl and COS. Nature *282,* 253–56.

Crutzen, P., 1979. The role of NO and NO_2 in the chemistry of the troposphere and stratosphere. Ann. Rev. Earth. Planet. Sci. *7,* 443–72.

Crutzen, P. J., M. J. Coffey, A. C. Delaney, R. Lueb, W. G. Mankin, W. Pollock, W. Seiler, A. Wartburgand, and P. Z. Zimmerman, 1984. Observations of air compositions in Brazil between the equator and 20°S during the dry season. Acta Amazonica (in press).

Dunnett, J. S., 1983. Ozone named as culprit. Nature *301,* 275.

Ehhalt, D. H., 1974. The cycle of atmospheric methane. Tellus *26,* 58–70.

Ehhalt, D. H. and Schmidt, 1978. Sources and sinks of atmospheric methane. Pure and Appl. Geophys. *116,* 452–64.

Ehhalt, D. H., R. J. Zander, and R. A. Lamontagne, 1983. On the temporal increase of tropospheric CH_4. J. Geosphys. Res. *88,* 842–46.

Evans, G., P. Finkelstein, B. Martin, N. Possiel, and M. Graves, 1983. Ozone measurements from a network of remote sites. J. Air Poll. Control Assoc. *33,* 291–96.

Fishman, J. 1981. The distribution of NO_x and the production of ozone: ozone by S. C. Livetal. J. Geophys. Res. *86,* 12,161–164.

Fraser, P. J., M. A. K. Khalil, R. A. Rasmussen, and A. J. Crawford, 1981. Trends in atmospheric methane in the Southern Hemisphere. Geophysical Research Letters *8,* 1,063–66.

Galbally, I. and C. R. Ray, 1978. Loss of fixed nitrogen from soils by nitric oxide exhalation. Nature *275,* 134–35.

Graedel, T. E. and J. E. McRae, 1982. Total organic component data: a study of urban atmospheric patterns and trends. Atmos. Environ. *16,* 1,119–132.

Heck, W. W., O. C. Taylor, R. Adams, G. Bingham, J. Miller, E. Preston, and L. Weinstein, 1982. Assessment of crop loss from ozone. J. Air Poll. Control Assoc. *32,* 353–61.

Hov, O., E. Hesstvedt, and I. S. A. Isaksen, 1978. Long range transport of tropospheric ozone. Nature *273,* 341–44.

Huebert, B. J. and A. L. Lazrus, 1980. Tropospheric gas phase and particulate nitrate measurements. J. Geophys. Res. *85,* 7,322–328.

Isaksen, I. S. A., O. Hov, and E. Hesstvedt, 1978. Ozone generation over rural areas. Environ. Sci. and Tech. *12,* 1,279–284.

Khalil, M. A. K. and R. A. Rasmussen, 1983. Sources, sinks and seasonal cycles of atmospheric methane. J. Geophys. Res. *88,* 5,131–144.

Khalil, M. A. K. and R. A. Rasmussen, 1984. Carbon monoxide in the earth's atmosphere: measuring trend. Science *224,* 54–56.

Lacis, A., J. Hansen, P. Lee, T. Mitchell, and S. Lebedeff, 1981. Greenhouse effect of trace gases 1970–1980. Geophys. Res. Lett. *8,* 1,035–38.

Levy, H., II, 1971. Normal atmosphere: Large radical and formeldehyde concentrations predicted. Science, *173,* 141–43.

Lipschultz, F., O. C. Zafirion, S. C. Wofsy, M. B. McElroy, F. W. Valois, and S. W. Watson, 1981. Production of NO and N_2O by soil nitrifying bacteria. Nature *294,* 641–44.

Logan, J. A., M. J. Prather, S. C. Wofsy, and M. B. McElroy, 1981. Tropospheric chemistry: a global perspective. J. Geophys. Res. *86,* 7,210–254.

Logan, J. A., 1983. Nitrogen oxides in the troposphere: global and regional budgets. J. Geophys. Res. *88,* 10,785–807.

Logan, J. A., M. J. Prather, S. C. Wofsy, M. B. McElroy, 1981. Tropospheric chemistry: a global perspective. J. Geophys. Res. *86,* 7,210–254.

Logan, J. A., M. J. Prather, S. C. Wofsy, and M. B. McElroy, 1978. Atmospheric chemistry: response to human influence. Phil. Trans. R. Soc. London, Ser. A, *290,* 187–234.

Logan, J. A., 1984. Surface-level ozone perturbations in rural and remote areas due to long-range transport of pollutants. (submitted, 1984).

McConnell, J. C., M. B. McElroy, and S. C. Wofsy, 1971. Natural sources of atmospheric CO. Nature *233,* 187–88.

McElroy, M. B., 1976. Chemical processes in the solar system: a kinetic perspective. In *MTP International Review of Chemical Kinetics,* ed. D. R. Herschbach, Butterworths, London, pp. 127–11.

McElroy, M. B., S. C. Wofsy, and Y. L. Yung, 1977. The nitrogen cycle: perturbations due to man and their impact on atmospheric N_2O and O_3. *Philosoph. Trans. Roy. Soc. (London) 277,* 159–81. National Academy of Sciences. Vapour phase organic pollutants. *National Research Council Panel on Vapor Phase Organic Pollutants,* NAS, Washington, D.C., 1976.

McElroy, M. B. (workshop chairman) et al., 1983. Global change: a biogeochemical perspective. Jet Propulsion Laboratory, publication 83–51, 33pp.

National Research Council, 1973. Biological impacts of increased intensities of solar ultraviolet reaction. Washington, D.C.: National Academy of Sciences.

National Research Council, 1975. Environmental impact of stratospheric flight: Biological and climatic effects of aircraft emissions in the stratosphere. Washington, D.C.: National Academy of Sciences.

National Research Council, 1976a. Halocarbons: Effects on stratospheric ozone. Washington, D.C.: National Academy of Sciences.

National Research Council, 1976b. Halocarbons, Environmental effects of chlorofluoromethane release. Washington, D.C.: National Academy of Sciences.

National Research Council, 1977. Medical and biologic effects of environmental pollutants: ozone and other photochemical oxidants. Washington, D.C.: National Academy Press.

National Research Council, 1979. Stratospheric ozone depletion by hydrocarbons: chemistry and transport. Washington, D.C.: National Academy of Sciences.

National Research Council, 1981. Atmosphere-biosphere interactions toward a better understanding of the consequences of fossil fuel combustion. Washington, D.C.

National Research Council, 1982. Causes and effects of stratospheric ozone reduction: an update. Washington, D.C.: National Academy Press.

National Research Council, 1983. Acid deposition: atmospheric processes in Eastern North America. Washington, D.C., 373 pp.

National Research Council, 1984a. Causes and effects of changes in stratospheric ozone: update, 1983. Washington, D.C., 254 pp.

National Research Council, 1984b. Global Tropospheric chemistry: a plan for action. Board on Atmospheric Sciences and Climate (in press, to be published September 1984).

Pierotti, D. and R. A. Rasmussen, 1976. Combustion as a source of nitrous oxide to the atmosphere. Geophys. Res. Lett. *3*, 265–67.

Possiel, N. C., J. F. Clarke, T. L. Clark, J. K. S. Ching, and E. L. Martinez, 1982. Recent EPA urban and regional scale oxidant field programs in the northeastern U.S. *Paper 82–24.1* presented at the 75th meeting of Air Pollution Control Association, New Orleans, LA, June.

Prinn, R. G., P. G. Simmonds, R. A. Rasmussen, R. D. Rosen, F. N. Alyea, C. A. Cardelino, A. J. Crawford, D. M. Cunnold, P. J. Fraser, and J. E. Lovelock, 1983. The atmospheric lifetime experiment. 1. Introduction, instrumentation, and overview. J. Geophys. Res. *88*, 8,353–367.

Raatz, W. E. and R. C. Schnell, 1984. Aerosol distributions and an arctic aerosol front during AGASP: Norwegian Arctic. Geophys. Res. Lett. *11*, 373–76.

Rudolf J. and D. H. Ehhalt, 1981. Measurements of C_2–C_5 hydrocarbons over the North Atlantic. J. Geosphys. Res. *86*, 11,959–964.

Samson, P. J. and K. W. Ragland, 1977. Ozone and visibility reduction in the midwest: evidence for large scale transport. J. Appl. Met. *16*, 1,101–106.

Schnell, R. C. and W. E. Raatz, 1984. Vertical and horizontal characteristics of Arctic Haze during AGASP: Alaskan Arctic. Geosphys. Res. Lett. *11*, 369–72.

Seiler, W., A. Holzapfel-Pschorn, R. Conrad, and D. Scharffe, 1984. Methane emissions from rice paddies. J. Atmos. Chem. *1*, 241–68.

Spicer, C. W., 1982. Nitrogen oxide reactions in the urban plume of Boston. Science *215*, 1,095–98.

Spicer, C. W., 1977. The fate of nitrogen oxides in the atmosphere. Adv. in Env. Sci. Tech. *7,* 163–61.

Spicer, C. W., D. W. Joseph, P. R. Sticksel, and G. F. Ward, 1979. Ozone sources and transport in the northeastern United States. Env. Sci. and Tech. *13,* 975–85.

Spicer, C. W., J. R. Koetz, G. W. Keigley, G. M. Sverdrup, and G. F. Ward, 1982a. Nitrogen oxide reactions within urban plumes transported over the ocean. *Contract No. 68–02–2957,* Draft Report for EPA, December.

Spicer, C. W., P. R. Sticksel, G. M. Sverdrup, A. J. Alkezweeny, and W. E. Davis, 1982b. Ozone and NO_x plumes from cities in the northeast corridor. *Paper 82–24.2* presented at the Air Pollution Control Association Meeting.

Sze, N. D., 1978. Anthropogenic CO emissions: Implications for the atmospheric CO–OH–CH_4 cycle. Science *195,* 673–75.

Vaughan, W. M., M. Chan, B. Cantrell, and F. Pooler, 1982. A study of persistent elevated pollution episodes in the northeastern United States. Bull. Am. Met. Soc. *63,* 258–66.

Vukovich, F. M., W. D. Bach, B. W. Crissman, and W. J. King, 1977. On the relationship between high ozone in the rural surface layer and high pressure systems. Atmos. Env. *11,* 967–83.

Warmbt, W., 1979. Krgebisse langjahriger Messungen des bodennahen Ozono in der DDR. Zeit schrift fur MEtcorology (Potsdeim), *29,* 24–31.

Weiss, R. F. and H. Craig, 1976. Production of nitrous oxide by combustion. Geosphys. Res. Lett. *3,* 751–53.

Weiss, R. F., 1981. The temporal and spatial distribution of tropospheric nitrous oxide. J. Geophys. Res. *86,* 7,185–195.

Wolff, G. T. and P. J. Lioy, 1980. Development of an ozone river associated with synoptic scale episodes in the eastern United States. Env. Sci. and Tech. *14,* 1,257–261.

Wolff, G. T., P. J. Lioy, R. E. Meyers, R. T. Cedarwall, G. D. Wight, R. E. Pasceri, and R. S. Taylor, 1977a. Env. Sci. and Tech. *11,* 506–10.

Wolff, G. T., P. J. Lioy, G. D. Wight, R. E. Meyers, and R. T. Cederwall, 1977b. An investigation of long range transport of ozone across the midwestern and eastern United States. Atmos. Env. *11,* 797–802.

6
POLLUTANT IMPACTS ON COASTAL ECOSYSTEMS

John M. Teal

INTRODUCTION

The impacts of coastal pollution ranging from those on coastal wetlands to those on dump site 106 have been reviewed in the last few years in a series of publications summarizing effects on single ecosystems,[1] effects of single classes of pollutants,[2] and overviews for near shore environments in general.[3] I will not repeat the material here, but will draw upon those sources and upon other recent literature to illustrate points I believe are important in relating long-term studies to pollutant impacts. The coastal zone is critical since it is an important source of food and the area that first receives most human waste that enters the sea.

We conduct long-term studies in any environment to understand how ecosystems function and how the impacts of human activities affect those functionings. We must begin with a knowledge of the range of variations through which an ecosystem passes normally. Without that knowledge, we have a very difficult time assigning a cause to a small pollution impact because we could not distinguish it from natural variations. Obviously, we also need knowledge of the processes that control the natural variations we observe and how pollutants interfere with these processes. Depending on the nature of the variability, the properties in whose variation we are interested, and the specific processes involved, this knowledge is likely to be difficult and expensive to obtain. But without knowledge of natural variability, we are forever reduced to a scholastic type of argument about what has happened to the system, what the causes are, and what we ought to do about it.

Concerns for coastal pollution and its effects, and therefore, I would argue, for long-term research in this type of environment, are widespread. Agencies at all levels of government are responsible for decisions that can affect coastal systems. For example, coastal municipalities apply for discharge permits for sewage which must be granted or refused by the EPA. Agencies concerned with public health, fisheries, and recreation have interests in the consequences of the discharges proposed. The concerns are quite clear even though the data and conclusions drawn from the data upon which the decisions are based may not be. It is less clear how the decisions of agricultural agencies with regard to persistent pesticides applied to farms far from the ocean, or agencies that have responsibilities for erosion control, will affect the health of coastal oceans but that they do have an effect is also clear.[4]

The users of coastal areas may express themselves very vocally. Harvesters of fish and shellfish, whether they do it for a living or for recreation, are interested in the continued health of coastal ecosystems and are often quick to blame changes they see on pollution. Coastal industries looking for the least expensive way of disposing of their wastes tend to err in the opposite direction by assuming that any damage observed is a part of natural variation in the system or due to overharvesting and not a result of their activities. Industrial and agricultural industries more than a few miles distant from coastal waters rarely relate their activities to changes in coastal ecosystems.

Individual academics and scientists, who may also have recreational and conservationist interests in the coast and be users of pesticides for control of lawn or garden problems, are interested in understanding the nature of coastal ecosystems and their interactions with society. As a scientist I maintain that long-term research on coastal systems is or should be a concern of all the groups. Understanding of ecosystem function is an essential ingredient in achieving any satisfactory consensus about the use of coastal environments. Summaries of positions and conflicts with regard to the effects of pollutants on oceans and their use as dumping grounds were outlined by Goldberg and Kamlet.[5]

Point Source Versus Nonpoint Source Pollution

Point sources of pollutants can, in theory, be controlled and treated at either source or discharge point. But in a point source discharge from a sewage treatment system are pollutants that originate in single sources—such as metals from the wastes of electroplating plants—which can be handled at the source and not allowed to enter the sewerage system. There are also metals that originate from nonpoint sources within the sewerage district. For example, metals from household piping are usually present in sewage but there is no reasonable input control available. They must be either treated before discharge or discharged into the receiving environment.

Metals that remain in the discharge could be disposed of with enough dilution so that the concentrations in marine sediments do not get above that of natural deposition processes. This would involve good initial dispersal in the water followed by enough dilution with benign material in the sediments where the metals are deposited. The sediment dilution could either rely on natural sedimentation or the supply of diluting particulates in the discharge. Metals enter the oceans naturally so their presence at background concentrations should not usually be a problem. Problems with such a scheme involve achieving the necessary dilution which is affected by processes from initial dispersal to resuspension by biological and nonbiological processes. If the concentration in sediments is not reduced to near background levels, then responses of organisms, including changes in toxicity of elements such as mercury and lead due to microbial alkylation, can become important.[6]

Organics such as petroleum hydrocarbons are naturally discharged into the oceans both in seeps and through the weathering and river transport of oil-bearing rocks. Man-made organics are different in that there may be no natural input to the oceans and hence no preadaptation to their presence and potential effects. Some such as the more heavily chlorinated PCBs degrade very slowly, accumulate in coastal areas, and are long lasting in their effects. They are also accumulated by organisms to a greater extent in relation to their concentration in the sediments than are petroleum hydrocarbons.[7] For these organics, dilution as a treatment would have to be sufficient to produce no effect even over a long time. We do not yet have the knowledge to know what a "safe" level is; keeping these compounds at the lowest possible levels in discharges seems the most reasonable course of action.

Inorganic nitrogen and phosphorus compounds (nutrients for plant production) can be removed from point source discharges but usually this is not done because of its cost. I would argue, therefore, that for practical purposes, nutrients from any source should be considered as not readily amenable to technological control. Nutrients from wastewater sources will ordinarily enter coastal environments as do pollutants from nonpoint sources such as nutrients derived from agricultural runoff, pesticides used on land and washed into the sea (for example DDT and pre-emergent herbicides) and soil itself when it is eroded and appears as sediment loads in runoff.

Pesticides can be restricted in their use to minimize runoff but it is very difficult to believe that this will ever be completely effective. Agriculture as typically practiced in the United States depends on use of herbicides to minimize the necessity for cultivation, especially with low or no-tillage methods. The use is likely to increase especially if crops resistant to herbicides are developed.[8]

I believe that our greatest problems in the future will be with nonpoint source pollutants and the most difficult to control will be those associated with nutrient overload which leads to eutrophication. The problem will ordinarily be greatest in the most enclosed estuaries and probably absent in the most open coastal areas subject to sufficient dilution to keep levels low. Nutrients are a problem because of the difficulty and expense of controlling them even at a point source and the much greater problem associated with nonpoint sources.

Monitoring

Monitoring is necessary to detect unanticipated results of human activities early enough to have the best chance to repair damage before it becomes irreversible. There are two aspects of monitoring necessary when considering pollutant impacts: monitoring inputs and monitoring effects. A watch over what is being produced and will subsequently be discarded or otherwise placed into the environment, and of what is appearing in coastal ecosystems provides an indication of what are possible causes of changes in ecosystems. The realization that substances such as toxiphenes and hexachlorocyclohexanes should be watched for is one example of input monitoring.[9] Another is the recent report of the highly toxic organic compound, 4-nonylphenol, a degradation product of a widely used surfactant, in ''extraordinarily high'' concentrations in anaerobically digested sewage sludge.[10] The Mussel Watch Program, designed to use the edible mussel as a sentinel organism, is an example of the second type of input monitoring which gives an indication within limitations imposed by what compounds are detected, and what the mussels accumulate, of materials found in coastal ecosystems.[11]

The second aspect of monitoring is observing what is actually occurring within the ecosystems themselves. We cannot possibly anticipate every type of pollutant that will reach our coastal waters, and we cannot possibly, on the basis of present knowledge, predict the long-range effects of those pollutants with which we are familiar.[12] We must, therefore, examine the coastal systems themselves for changes in structure and function and try to relate these to pollution inputs.

Other Concerns

My list of reasons why long-term research in coastal ecosystems is important is, for the most part, a list of the impacts of society on these systems and the difficulty of understanding the consequences of these impacts without much better time series data than we now have. I add here some aspects of impacts on coastal systems that do not always, for a variety of reasons, come under the heading of pollution.

The most obvious impacts are the changes in sediment supply that result from changes in rates of erosion and river manipulations. Increases in erosion could have many biological effects in coastal waters, for example: lowering of productivity due to reduction in light penetration with consequences for benthic faunas and floras and wetland maintenance. Turbidity varies widely under natural conditions, and, again, only with careful long-term examination can we understand the consequences of our actions.

River management can change sediment loads and interfere with the ecology of organisms, e.g., anadromous species. Dams provide settling basins which remove sediments. Removal of old dams with filled basins suddenly provides increased sediment loading as the river cuts into the sediment pond. River diversions rechannel sediments, most dramatically in the case of the lower Mississippi, where lack of sediment is contributing to the drastic loss of Louisiana wetlands.[13]

According to some predictions sea level could rise between 1.5 and 2 meters by 2100.[14] The rise will be due to atmospheric pollution, not direct pollution of the coastal oceans, but will have a drastic effect there. Erosion will be increased, wetlands and barrier islands flooded. The natural processes by which wetlands maintain themselves by vertical growth and growing up over flooded uplands will be prevented by structures either designed to prevent flooding, such as levees, or by structures that will greatly reduce flooding regardless of their purpose, for example, railroad beds. The normal movements of barrier islands which now maintain their relative position by rolling-over into shallower water will be hindered by efforts to protect structures on them and waterways behind them, as well as the barriers. In both cases the shallow water ecosystems and the organisms which apparently depend upon them will be greatly affected. Some long series time data during the preliminary, slower stages of sea level rise would give us insight on how to adapt to coming changes, if not how to control them.

Finally, changes in amount and periodicity of water discharge from rivers, as those rivers are used increasingly for irrigation and power generation will change tidal circulation in the receiving estuaries and the distribution of sediments. These will again have long-term effects on production and distribution of coastal organisms.

PRESENT STATUS OF KNOWLEDGE

We have relatively little in the way of long-term data for coastal ecosystems. This lack results to a considerable extent from our lack of experience in the broadest sense with underwater environments. We have thousands of years of cultural experience with land environments. We have accumulated practical experience with terrestrial ecosystems through agriculture and forestry. There exists a long historical record with streams and ponds but mostly from looking down through the surface. Aside from the intertidal, even this relatively distant view has been mostly lacking with regard to the sea, except for fishery records which represent reality filtered both through the fishermen's nets and their frequent efforts to mislead authority about where they catch what. Only in the last few decades have we been able to look closely at marine organisms functioning in an undisturbed manner in their environment. SCUBA gives us access both for scientists and general observers. Research submarines and remotely controlled underwater devices give us access for observation and measurements into the deeper parts of the oceans. Satellites provide a way of surveying huge areas of the surface oceans rapidly for synoptic views never before possible. These technologies have allowed scientists to venture beyond laboratory experiments to experiments with the subtidal and open water parts of the oceans where most of the natural complexity of the systems can be included and considered.

Point Source Pollutants

For situations in which point sources have resulted in relatively high concentrations of pollutants, we have the best data on effects for any situation in the marine environment. Additions of organic matter, great enough to cause anoxia, kill all the resident fauna at the discharge site. As one moves away from the point at which the organic matter enters the system, one typically sees a maximum in abundance of animals, but the fauna is composed of only a few opportunistic, subsurface, nonselective deposit feeding species such as the polychaete worm *Capitella*. Further from the source of contamination, there is usually recovery in terms of species diversity, abundance, and function of the normal local community.[15]

Organic chemicals can cause tumors, impair growth and reproduction, and modify the genetic material within organisms. Metals can also be toxic, but organisms may have better mechanisms for dealing with them.[16] But even in the cases of heavy pollution in confined areas, many of the modifications to the system and its populations cannot be assigned to specific cause because so many pollutants are present together.

Experimental approaches using large enclosures (mesocosms) or partial enclosures are necessary to study in a controlled manner, the interactions between pollutants and the various parts of an ecosystem acting in concert. In nature pollutants virtually always act along with a great variety of other contaminants we have introduced into the system along with other factors such as harvesting. Only when a pollutant is present in relatively high concentration with respect to its dose-response curve, can we unequivocally ascertain its effect on the ecosystem. But with such effects, we are dealing with acute, not long-term effects. By using experimental techniques with controls, we can begin to understand the consequences of individual compounds and classes of compounds. Laboratory experiments can deal with individual compounds and individual species but ordinarily lack the population interactions characteristic of ecosystems.

This mesocosm-scale experimental approach is particularly important for research with plankton in which field studies are made extremely difficult by the variability in time and space to which they are subject. Studies with planktonic systems enclosed in large plastic bags (for example, Contained Ecosystem Pollution Experiment or CEPEX)[17] have provided data on the effects of hydrocarbons and metals on plankton populations. But even these experiments are difficult to interpret unequivocally because of the high variability both in the starting populations in the containers and their subsequent behavior. Though solely planktonic systems are difficult to experiment with, better results have been obtained with combined planktonic-benthic systems such as the Marine Ecosystem Research Laboratory (MERL) tanks at the University of Rhode Island which successfully track conditions and populations in Narragansett Bay. Interactions between populations are more easily seen in mesocosm experiments as illustrated by the stimulation in phytoplankton production in a pollution experiment with fuel oil caused by inhibition of grazing by zooplankton and benthic filter feeders, rather than any direct effect on the phytoplankton themselves.[18]

The mesocosm approach to ecosystem study is limited by its inability to include the largest features of the systems under study. Large predators, for example, fishes over a few centimeters length, require larger tanks than the 1.8 m diameter by 5 m deep MERL tanks. It is not possible to imitate the full range of circulation and mixing events in the tanks, especially the rare severe events such as storms. Probably the next step will require multiple, permanent tanks with volumes on the order of 10^{3-4} m^3.

Recovery processes are also susceptible to experimental study in mesocosms. After a MERL experiment that compared the effects of different degrees of pollution in sediments on an estuarine ecosystem, researchers predicted "that removing pollutant inputs will, after a short time, result in a relatively clean and biologically normal water column."[19] Though nutrient concentrations and production rates remained slightly higher for the first year in the systems with the highest level of pollution in the sediments as compared with the controls, both were much lower than in the ecosystem from which the polluted sediments were obtained.[19] This indicates that even the consequences of eutrophication in relatively enclosed coastal water bodies would disappear rather rapidly if the inputs were stopped. The sediments would remain polluted for a long time and would continue to be a source of pollutants to the water column. Hunt and Smith[20] found export of copper, lead, and cadmium to the water column from the polluted sediments in the above experiment. The release of metals was probably enhanced both by the increase in redox due to the reduction in organic input and to the increased irrigation of the sediments by the enhanced animal populations. It would have required about 44 years for the top centimeter of their highly polluted sediment to lose enough copper (84%) to get down to the level of the only moderately polluted Narragansett Bay, but the populations were considered to be equivalent to those in the controls within five months.[19] If it were possible to clean up the input to the polluted site, the natural deposition of clean sediment on top of the polluted sediment would reduce the loss of pollutants to the water column.

Nonpoint Source Pollutants

The general addition of nutrients to coastal oceans enhances production and can be relatively harmless. In South San Francisco Bay, which is a very shallow water body, the water column production is controlled by benthic consumers.[21] Enhanced production by phytoplankton supports an enhanced benthic population (mostly of introduced organisms) but does not result in deleterious conditions such as lack of oxygen in the water. Marine wetlands are capable of handling large additions of nutrients without apparent harmful effects.[22]

In contrast, large inputs of nutrients have apparently been responsible along with summer stratification for the increasing anoxia found in the bottom waters of Chesapeake Bay, for example. The condition has been increasing since the 1930s but has become critical in recent years. There are widespread consequences in fisheries yields and in species available.[23] Anoxia can also lead to changes in metal remobilization. Some metals, such as lead, are readily bound in reduced sediments as sulfides, while others are mobilized under reducing conditions[24] which in turn lead to increases in numbers of benthic polychaetes.

Nixon has summarized the existing data for the best studied estuaries in the United States and finds that there is a correlation between nutrient loading to estuaries and mean annual nutrient concentration in estuaries, and between nutrient loading and chlorophyll-a concentration[25] that is, there does seem to be a connection between nutrient input and eutrophication even though it is not very soundly documented.

Submerged vascular plants have been declining in Chesapeake Bay since the 1960s because of declining light levels due to increased sediment load and increased plant nutrients. Suspended sediments absorb light directly. Nutrients stimulate production by phytoplankton and epiphytes growing on the submerged vegetation which in turn capture light that would otherwise reach the vascular plants.[26] Submerged plants along with wetlands reduce erosion by absorbing wave energy, provide nursery ground for the young of valuable fish and shellfish, and probably contribute to their productivity, so their decline is significant. Studies in ponds also showed that the total productivity of the system was reduced by high concentrations of nutrients since the shading effects outweighed the direct stimulation of growth when all components of the system were considered together.[27] Herbicides in runoff from agricultural land has been implicated in the macrophyte decline but because the major pollutant, atrazine, has a very limited lifetime in estuaries it has probably been significant only in confined areas receiving concentrated runoff[26] though the effect can become more serious if use and runoff increases significantly. A pollution caused decrease in macrophytes in a part of Apalachee Bay, Florida, led to an overall reduction in productivity and to changes in the fish communities, though in that instance most of the pollution was from a point source.[28]

Even without drastic changes in ecosystems such as onset of anoxia in the overlying waters, organic enrichment produces changes in benthic populations. In the extreme case short of anoxia, the normally resident benthic species of the New York Bight were replaced by opportunists unsuitable as food for predators such as fish.[29]

To give just one example of the effects of changes in river flow, along the Mediterranean coast of France, sea grass beds have been greatly reduced. Part of the effect is due to changes in circulation and deposition of fine sediments resulting from harnessing of the normal flow of the Rhône for hydroelectric generation and the consequent reduction in flood discharges. The other factor in the decline of the beds is, as in the Chesapeake, a reduction in light due to eutrophication from sewage pollution.[30]

RECOMMENDATIONS

There are many gaps in our knowledge of how coastal ecosystems function. We are only at the beginning of having enough understanding of the best studied coastal systems—salt marshes and estuaries—to be able to contribute to their management effectively. But we almost completely lack long-term time series data on principal variables that have been shown by process-oriented experiments to be significant for control of marine ecosystem structure and function with the situation much worse for estuaries than marshes.

Aquatic ecosystems must be examined regularly and in some detail if we are ever to untangle the complexities of the processes that govern how they function and the impacts of pollutants upon them. Ideally this examination should be in the context of a long-term research program that not only maintains some level of general monitoring, but also accompanies this with research programs both in the field and the laboratory to unravel the processes that control the system. We must have long time-series data if we are ever to distinguish between natural variations and pollution-induced changes. In the case of the Chesapeake Bay anoxia, Officer et al.[23] list six important processes they believe need further study for the proper analysis of the anoxia events. I would point out that they can make as definitive an analysis as they have only because they have at least some data over nearly 50 years.[31] But even for the Chesapeake there is a real dearth of volume-integrated data even for nutrients and production over any significant time period.[25]

Ecosystem monitoring must not be neglected just because it is difficult to separate the signal from the noise,[32] for it is only by looking at nature in its entirety that it is possible to detect the multitude of unexpected interactions and higher level effects that may occur as a result of modifications of natural systems that can result from the introduction of pollutants. Lewis[32] correctly suggests that more effort should be devoted to study in the area between laboratory effects in an artificial setting on one hand and community ecology with the entire system on the other.

I would repeat the plea I have made before that scientists and managers should pay more attention to the myriad of observers present such as fishermen, bird watchers, SCUBA divers, and shellfish harvesters who together spend much more time looking at coastal ecosystems than scientists do. They will frequently be the first to detect something going wrong. They will often be wrong in assigning cause and effect to what they have observed. Fishermen, for example, are the last to believe that their own fishing has any connection with an observed decline in fish stocks. Nevertheless, they may well be the first to know that stocks have changed significantly, and their definition of "significance" (catch per effort has declined enough for them to notice economically or recreationally) is a definition of great importance to managers and political representatives. The effect of DDT on bird reproduction is another often cited example. Once the effect was noticed, scientists discovered what to look for in the birds, managers and politicians decided how to control it, and, as a result, the peregrine falcon and other avian accumulators of DDT still exist. Another example is found in the spread and documenting of anoxic conditions in Chesapeake Bay by commercial fishermen.

To return to the subject of my first recommendation, there is a great need for support of research on a long-term basis for some coastal ecosystems where a combination of regular examination of the system for building up good time-series data can be coupled with continued research into the processes which control how the ecosystem functions. This requires a nearby university or similar facility equipped in both personnel and scientific apparatus to pursue such research. Long-term data sets require good quality control and careful design for scientific and statistical reliability. They need alert, critical scientists who can judge whether an odd datum is just bad, or should suggest a previously unsuspected process that deserves investigation. Ideally there would also be nearby management organizations so that the researchers would have regular contact with the users of their information as far as the control of effects of pollution are concerned. I also firmly believe that a social science research facility should be coupled with the above to make the best use of all the associated efforts and to help in the understanding of how knowledge of one aspect of a natural system is incorporated into its management. I find nothing quite so depressing as the common though usually unstated assumption on the part of natural scientists that the "truth" as revealed by their research will naturally and almost immediately result in regulations incorporating what they see as the consequences of their results.

I believe scientific results fully justify the support of facilities such as the MERL at the University of Rhode Island. Probably support should come in terms of the basic knowledge gained there, that is from NSF, rather than from the specific knowledge about specific pollutants. The latter is supported by the mission-oriented agencies, which as far as pollution is concerned are far too subject to the vagaries of politics to provide useful long-term support.

NOTES

1. G. F. Mayer, (ed.), *Ecological Stress and the New York Bight: Science and Management* (Estuarine Research Federation, Columbia, S.C. 1982). J. M. Teal, "Long-term Studies of Pollution in One Salt Marsh" in *Productivity, Pollution and Policy in the Coastal Zone,* Proc. Conf. in Rio Grande do Sul, Brasil, in press.

2. D. F. Boesch et al., *Assessment of Long-term Effects of OCS Development* (Federal Interagency Committee on Ocean Pollution Research, Development and Monitoring), in press.

3. B. H. Ketchum et al., *Wastes in the Ocean* (Wiley Interscience, New York, 1984).

4. E. Cronin, "Chesapeake Bay: A Case Study" in *Planning, Pollution, and Policy in the Coastal Zone,* W. Kirby-Smith, (ed.), Proc. Conf. in Rio Grande do Sul, Brasil, in press.

5. E. D. Goldberg, "The Oceans as Waste Space: The Argument," *Oceanus* 24 (1981):2–9. K. S. Kamlet, "The Oceans as Waste Space: The Rebuttal," *Oceanus* 24 (1981):10–17.

6. E. D. Goldberg, "Disposal of Industrial and Domestic Wastes. Land and Sea Alternatives" (Washington, D.C.: National Research Council National Academy of Sciences Press, 1984).

7. J. W. Farrington et al., "Experimental Studies of Hydrocarbons in Benthic Animals and Sediments" in *Pollution Transfer Processes,* M. Waldichuk, (ed.), International Association for the Physical Science of the Oceans, in press.

8. Constance Matthiessen and Howard Kohn, "In Search of the Perfect Tomato," *The Nation* 239 (1984):13–15.

9. E. D. Goldberg, "Can the Oceans Be Protected?" *Canadian Journal of Fisheries and Aquatic Sciences, (Supplement No. 2)* 40 (1983):349–535.

10. W. Giger et al., "4-Nonylphenol in Sewage Sludge: Accumulation of Toxic Metabolites from Nonionic Surfactants," *Science* 225 (1984):623–25.

11. National Academy of Sciences, *The International Mussel Watch,* report of a workshop sponsored by the Environmental Studies Board, Commission on Natural Resources, National Research Council (Washington, D.C.: NAS., 1984), 77pp.

12. S. W. Nixon et al., "Eutrophication of a Coastal Marine Ecosystem—An Experimental Study Using the MERL Microcosms" in *Flows of Energy and Materials in Marine Ecosystems,* M. J. R. Fasham, (ed.), (NY: Plenum, 1984).

13. R. H. Baumann et al., "Mississippi Deltaic Wetland Survival: Sedimentation Versus Coastal Submergence," *Science* 224 (1984):1,093–95.

14. J. S. Hoffman et al., "Projecting Future Sea Level Rise" (Washington, D.C.: U.S. Environmental Protection Agency, EPA 230–09–007, 1983).

15. T. H. Pearson and R. Rosenberg, "Macrobenthic Succession in Relation to Organic Enrichment and Pollution of the Marine Environment," *Oceanography and Marine Biology, An Annual Review* 16 (1978):229–311.

16. J. M. Capuzzo et al., "The Impact of Waste Disposal in Nearshore Environments" in *Wastes in the Ocean, Vol 6., Nearshore Waste Disposal,* B. H. Ketchum, J. M. Capuzzo, W. V. Burt, I. W. Duedall, P. K. Park, and D. R. Kester, (eds.), (New York, NY: Wiley Interscience, 1984).

17. G. P. Grice and M. R. Reeve, (eds.), *Marine Mesocosms* (New York: Springer, 1982).

18. R. Elmgren et al., "Trophic Interactions in Experimental Marine Ecosystems Perturbed by Oil" in *Microcosms in Ecological Research,* J. P. Giesey, (ed.), (Washington D.C.: U.S. Department of Energy, 1980).

19. C. A. Oviatt et al., "Recovery of a Polluted Estuarine System: A Mesocosm Experiment" *Marine Ecology Progress Series* 9 (1984):203–17.

20. C. D. Hunt and D. L. Smith, "Remobilization of Metals from Polluted Marine Sediments" *Canadian Journal of Fisheries and Aquatic Sciences, (Supplement No. 2)* 40 (1983):132–42.

21. J. E. Cloern, "Does the Benthos Control Phytoplankton Biomass in South San Francisco

Bay?" *Marine Ecology Progress Series* 9 (1984):191–202; C. B. Officer et al., "Benthic Filter Feeding: A Natural Eutrophic Control," *Marine Ecology Progress Series* 9 (1984):203–10; Nixon et al. also found in MERL experiments that nutrient additions stimulated planktonic diatom production.

22. J. M. Teal, "Long-term Studies of Pollution in One Salt Marsh" in *Planning, Pollution, and Policy in the Coastal Zone,* W. Kirby-Smith, (ed.), Proc. Conf. in Rio Grande do Sul, Brasil, in press.

23. C. B. Officer et al., "Chesapeake Bay Anoxia: Origin, Development, and Significance," *Science* 223 (1984):22–27.

24. A. E. Giblin et al., "Heavy Metal Uptake in a New England Salt Marsh," *American Journal of Botany* 67 (1980):1059–68; G. A. Jackson, "Sludge Disposal in Southern California Basins," *Environmental Science and Technology* 16 (1979):746–57.

25. S. W. Nixon, "Estuarine Ecology—A Comparative and Experimental Analysis using 14 Estuaries and the MERL Microcosms" (Final Report to the U.S. Environmental Protection Agency, Chesapeake Bay Program under Grant No. X–003259–01, 1984).

26. W. M. Kemp et al., "The Decline of Submerged Aquatic Vegetation in Chesapeake Bay: A Summary of Results Concerning Possible Causes," *Marine Technology Society Journal* 17 (1983):78–89; R. L. Wetzel and P. A. Penhale, "Production Ecology of Seagrass Communities in the Lower Chesapeake Bay," *Marine Society Technology Journal* 17 (1983):22–31.

27. R. R. Twilley et al., "Nutrient Enrichment of Estuarine Submersed Vascuar Plant Communities: 1. Algal Growth and Effects on Production of Plants and Associated Communities," Mar. Ecol. Prog. Ser. 23 (1985):179–92.

28. R. J. Livingston, "Trophic Responses of Fishes to Habitat Variability in Coastal Seagrass Systems," *Ecology* 65 (1984):1,258–275.

29. D. F. Boesch, "Ecosystem Consequences of Alterations of Benthic Community Structure and Function in the New York Bight Region" in *Ecological Stress and the New York Bight: Science and Management,* G. F. Mayer, (ed.), (Columbia, S.C.: Estuarine Research Federation, 1984).

30. J. M. Peres and J. Picard, "Causes de la Raréfaction et de la Disparition des Herbiers de *Posidonia oceanica* sur les Côtes Francaises de la Méditerranée," *Aquatic Botany*. 1 (1975):133–39.

31. Long-term research projects in coastal environments are or have been supported at present by the Ecosystems Program of NSF, by the Outer Continental Shelf Program of the Minerals Management Service, by NOAA, and EPA. EPA was responsible for a major portion of the support of MERL during the initial and critical years.

32. J. R. Lewis, "Options and Problems in Environmental Management and Evaluation," *Helgolander Meeresunters* 33 (1980):452–66.

ADDITIONAL REFERENCES

Means, J. C., R. D. Wijayaratne, and W. R. Boynton. 1983. "Fate and Transport of Selected Pesticides in Estuarine Environments." *Canadian Journal of Fisheries and Aquatic Sciences (Supplement No. 2)* 40:337–45.

Valiela, I. 1984. "Nitrogen in Salt Marsh Ecosystems." In *Nitrogen in the Marine Environment,* E. J. Carpenter and D. G. Capone (eds.) pp. 649–78. Academic Press, New York.

BIBLIOGRAPHY

Baumann, R. H., J. W. Day, Jr., and C. A. Miller. 1984. "Mississippi Deltaic Wetland Survival: Sedimentation Versus Coastal Submergence." *Science* 224:1093–1095.

Boesch, D. F. 1982. "Ecosystem Consequences of Alterations of Benthic Community

Structure and Function in the New York Bight Region." In *Ecological Stress and the New York Bight: Science and Management*, G. F. Mayer, (ed.), Estuarine Research Federation, Columbia, S.C.

Boesch, D. F. et al. 1984. *Assessment of Long-term Effects of OCS Development*. Federal Interagency Committee on Ocean Pollution.

Capuzzo, J. M., W. V. Burt, I. W. Duedall, P. K. Park, and D. R. Kester. 1984. "The Impact of Waste Disposal in Nearshore Environments." In *Wastes in the Ocean, Vol. 6., Nearshore Waste Disposal*, B. H. Ketchum, J. M. Capuzzo, W. V. Burt, I. W. Duedall, P. K. Park, and D. R. Kester (eds.). Wiley Interscience, New York.

Cloern, J. E. 1984. "Does the Benthos Control Phytoplankton Biomass in South San Francisco Bay?" *Marine Ecology Progress Series* 9:191–202.

Cronin, E. 1984. "Chesapeake Bay: A Case History." In *Planning, Pollution, and Policy in the Coastal Zone*, W. Kirby-Smith, (ed.). Proc. Conf. in Rio Grande do Sul, Brasil.

Elmgren, R., G. A. Vargo, J. F. Grassle, J. P. Grassle, D. R. Heinle, G. Langlolis, and S. L. Vargo. 1980. "Trophic Interactions in Experimental Marine Ecosystems Perturbed by Oil." In *Microcosms in Ecological Research*, J. P. Giesey, (ed.). U.S. Department of Energy, Washington, D.C.

Farrington, J. W., J. M. Teal, and B. W. Tripp. 1981. "Experimental Studies of Hydrocarbons in Benthic Animals and Sediments." In *Pollution Transfer Processes*, M. Waldichuk, (ed.). International Association for the Physical Science of the Oceans. In press.

Giblin, A. E., A. Bourg, I. Valiela, and J. M. Teal. 1980. "Heavy Metal Uptake in a New England Salt Marsh." *American Journal of Botany* 67:1059–68.

Giger, W., P. H. Brunner, and C. Schaffner. 1984. "4-Nonylphenol in Sewage Sludge: Accumulation of Toxic Metabolites from Nonionic Surfactants." *Science* 225:623–25.

Grice, G. P and M. R Reeve, (eds.). 1982. *Marine Mesocosms*. Springer, New York.

Goldberg, E. D. 1981. "The Oceans as Waste Space: The Argument." *Oceanus* 24:2–9.

Goldberg, E. D. 1983. "Can the Oceans Be Protected?" *Canadian Journal of Fisheries and Aquatic Sciences, (Supplement No. 2)* 40:349–535.

Goldberg, E. D. 1984. "Land, Sea and Air Options for the Disposal of Industrial and Domestic Wastes." National Research Council, National Academy of Sciences Press, Washington, D.C.

Hoffman, J. S., D. Keyes, and J. G. Titus. 1983. "Projecting Future Sea Level Rise." U.S. Environmental Protection Agency. EPA 230–09–007. Washington, D.C.

Hunt, C. D. and D. L. Smith. 1983. "Remobilization of Metals from Polluted Marine Sediments." *Canadian Journal of Fisheries and Aquatic Sciences, (Supplement No. 2)* 40:132–42.

Jackson, G. A. 1979. "Sludge Disposal in Southern California Basins." *Environmental Science and Technology* 16:746–57.

Kamlet, K. S. 1981. "The Oceans as Waste Space: The Rebuttal." *Oceanus* 24:10–17.

Kemp, W. M., W. R. Boynton, R. R. Twilley, J. C. Stevenson, and J. C. Means. 1983. "The Decline of Submerged Aquatic Vegetation in Chesapeake Bay: A Summary of Results Concerning Possible Causes." *Marine Technology Society Journal* 17:78–89.

Ketchum, B. H., J. M. Capuzzo, W. V. Burt, I. W. Duedall, P. K. Park, and D. R. Kester, (eds.). 1984. *Wastes in the Ocean*. Wiley Interscience, New York.

Lewis, J. R. 1980. "Options and Problems in Environmental Management and Evaluation." *Helgolander Meeresunters* 33:452–66.

Livingston, R. J. 1984. "Trophic Responses of Fishes to Habitat Variability in Coastal Seagrass Systems." *Ecology* 65:1258–275.

Mathiessen, C. and H. Kohn. 1984. "In Search of the Perfect Tomato." *The Nation* 239:1984.

Mayer, G. F., (ed.) 1982. *Ecological Stress and the New York Bight: Science and Management*. Estuarine Research Federation, Columbia, S.C.

Means, J. C., R. D. Wijayaratne, and W. R. Boynton. 1983. "Fate and Transport of Selected Pesticides in Estuarine Environments." *Canadian Journal of Fisheries and Aquatic Sciences (Supplement No. 2)* 40:337–45.

National Academy of Sciences. 1980. "The International Mussel Watch." Report of a workshop sponsored by the Environmental Studies Board, Commission on Natural Resources, National Research Council. 77pp. Washington, D.C.

Nixon, S. W., M. E. Q. Pilson, C. A. Oviatt, P. Donaghay, B. Sullivan, B. Seitzinger, D. Rudnick, and J. Frithsen. 1984. "Eutrophication of a Coastal Marine Ecosystem—An Experimental Study Using the MERL Microcosms. In *Flows of Energy and Materials in Marine Ecosystems*, M. J. R. Fasham, (Ed.). Plenum, NY.

Nixon, S. W. 1983. "Estuarine Ecology—A Comparative and Experimental Analysis using 14 Estuaries and the MERL Microcosms." Final Report to the U.S. Environmental Protection Agency, Chesapeake Bay Program under Grant No. X–003259–01.

Officer, C. B., R. B. Biggs, J. L. Taft, L. E. Cronin, M. A. Tyler, and W. R Boynton. 1984. "Chesapeake Bay Anoxia: Origin, Development, and Significance." *Science* 223:22–27.

Officer, C. B., T. J. Smayda, and R. Mann. 1982. "Benthic Filter Feeding: A Natural Eutrophic Control." *Marine Ecology Progress Series* 9:203–10.

Oviatt, C. A., M. E. Q. Pilson, S. W. Nixon, J. B. Frithsen, D. T. Rudnick, J. R. Kelly, J. F. Grassle, and J. P. Grassle. 1984. "Recovery of a Polluted Estuarine System: A Mesocosm Experiment." *Marine Ecology Progress Series* 16:203–17.

Pearson, T. H. and R. Rosenberg. 1978. "Macrobenthic Succession in Relation to Organic Enrichment and Pollution of the Marine Environment." *Oceanography and Marine Biology, An Annual Review* 16:229–311.

Peres, J. M. and J. Picard. 1975. "Causes de la Raréfaction et de la Disparition des Herbiers de *Posidonia oceanica* sur les Côtes Francaises de la Méditerranée." *Aquatic Botany* 1:133–39.

Teal, J. M. 1984. "Long-term Studies of Pollution in One Salt Marsh." In *Planning, Pollution, and Policy in the Coastal Zone*, W. Kirby-Smith (ed.). Proc. Conf. in Rio Grande do Sul, Brasil. In press.

Twilley, R. R., W. M. Kemp, K. W. Staver, J. C. Stevenson, and W. R. Boynton. 1985. Nutrient Enrichment of Estuarine Submersed Vascuar Plants and Communities: 1. Algal Growth and Effects on Production of Plants and Associated Communities. Mar. Ecol. Prog. Ser. 23:179–92.

Valiela, I. 1984. "Nitrogen in Salt Marsh Ecosystems." In *Nitrogen in the Marine Environment*, E. J. Carpenter and D. G. Capone, (eds.), pp. 649–78. Academic Press, New York.

Wetzel R. L. and P. A. Penhale. 1983. "Production Ecology of Seagrass Communities in the Lower Chesapeake Bay." *Marine Society Technology Journal* 17:22–31.

7

SCIENTIFIC, LEGISLATIVE, AND ADMINISTRATIVE CONSTRAINTS TO MULTIMEDIA CONTROL OF TOXIC SUBSTANCES AND HAZARDOUS WASTES

Leslie Sue Ritts
and Roger C. Dower

INTRODUCTION: A HYPOTHETICAL

Imagine a chemical named benxocartite. A petroleum derivative, benxocartite (BOC) is one of a family of compounds with production levels of approximately 1 million pounds per year. BOC's primary application is as an active ingredient in the pesticide Lesritts, used on corn. Its other chemical family members do not share its useful pesticide characteristics. On the other hand, they have several profitable industrial applications as intermediaries (bonding agents) in industrial glues. All BOC family members begin life as benxo (BO), produced by a small number of large chemical manufacturers who sell the chemical to specialty manufacturers where BO is transformed into BOCs and its other derivatives.

Epidemiological studies of both male and female workers exposed to BO have recently demonstrated an increased incidence of liver cancer and statistically high incidences of sterility. Very little else, however, is known concerning environmental exposures or health effects of BOC or the other members of the BO family. There are no known studies concerning the toxicological effects of BO at the current low levels of exposure. Nor are there protocols for studying the effects of BOC in the environment or their absorption by organisms in the food chain. Monitoring is not presently capable of distinguishing the various derivative BO family members.

BOs have recently been noted in monitoring data at a number of public water systems in relatively low concentrations. Some of these are drawn from large freshwater lakes. More recently, trace amounts of BO have also been found in certain groundwater aquifers near plants that manufacture Lesritts. The chemical's presence appears to be related to residuals from on-site industrial and off-site municipal landfills where surplus Lesritts is disposed of, but may also be associated with atmospheric transport of the organic chemical from stack emissions.

A just published investigation suggests that BOC may be associated with a wider range of acute health responses in field workers, particularly when used in combination with another common herbicide, Rower (RR). Further, farming practices most often associated with the major crop use of Lesritts have been shown to lead to excessive soil erosion in the southeastern portion of the United States. BO, however, has not been identified as a contaminant in underground aquifers where the pesticide is widely used.

Discussion of Hypothetical

The hypothetical described above provides a basis for several observations concerning scientific, legislative, and administrative constraints to integrating the control of toxic substances throughout their production, use, and disposal. So as not to become mired in some of the scientific issues that pervade the evaluation of any potentially toxic substance at the beginning of this discussion (for instance, questions about exposure thresholds, validity of epidemiological studies, and interpolation models—to name but a few), legislative and administrative constraints in addressing a potentially hazardous chemical like BOC will be discussed first. ''Science policy'' as it emanates from the legislation and administration of environmental laws will then be discussed.

When confronted with our hypothetical, the current regulatory structure is likely to result in an array of emission limits, maximum contaminant levels, effluent guidelines, manufacturing guidelines, and registration requirements directed toward one or more uses of BO and its derivatives like BOC. These policies for addressing BO and BOC in different media may or may not coincide with the most cost-effective or rational regulatory responses for addressing potentially hazardous human and environmental exposures to a toxic substance. For instance, precommercialization requirements may provide little screening of BO's behavior in the air or in the soil, their transformation in these media, or behavior in water. None of these schemes will probably directly address, for instance, the reactivity of trace amounts of vaporized BOC in the air, its reaction with other primary pollutants or the distance it travels before it is deposed on the soil. No present statutory scheme mandates an examination of how BOC reacts with the active ingredients of Rower, or their potential to leach into ground and surface waters. In much the same ways up until 1976, statutes ignored landfilled BOCs which were thought to be inert when disposed of in the soil. These regimes still ignore the hazards of the emissions that may be released into the atmosphere from disposal facilities as particulates.

In addition, each regulatory action affecting BO or BOC involves differing legal requirements affecting rule making and a host of other administrative and managerial procedures for examining and regulating compliance with rules that control BOC's manufacture, use, and disposal. Moreover, agency resources, tied-up in examining the possibility of regulating BOC, may result in other potentially more hazardous substances not receiving EPA's attention. Taken together, statutory and administrative constraints, together with failure to prioritize environmental risks involving the regulation of BO and BOC in the various media will almost certainly mean that attempts to control BOC will misallocate resources, at the least, and be ineffective at safeguarding the public's health and environmental resources, at the worst.

The recommended solution to all these problems is improved integration of our toxic and hazardous waste regulatory programs either across media, pollutant, or receptor by means of administrative or legislative reform. Several research projects have been conducted or are underway that document the nature and the extent of our failure to integrate the regulation of toxic substances. These suggest that the benefits from a more integrated approach to chemical management could be significant. At the same time, existing constraints to integration (with their implied costs) are also significant and must be weighed against the expected benefits of toxics integration.

The careful design of a national response to the problem posed by pollutants with multimedia effects requires that both the anticipated benefits and costs of reordering existing regulatory responses be carefully articulated. The remainder of this chapter will review some of the more significant legislative, administrative, and science-policy constraints to integrated regulatory programs, will describe current activities in each of these areas to make existing regulatory programs more consistent, and will present several research goals that the authors think could define more clearly the potential costs, as well as the benefits, of alternative treatments of toxic substance and hazardous waste problems.

LEGISLATIVE CONSTRAINTS

Statement of the Problem

There can be little doubt that the existence of eleven major environmental statutes is a major impediment to increasing the effectiveness of EPA in responding to the problems of toxics integration.[1] At least ten other major federal government departments or independent agencies also have authority for administering laws affecting toxic substances in the environment.[2]

The Conservation Foundation Mid-Decade State of the Environment Report notes that when vinyl chloride was recognized as a health hazard in the mid 1970s, five federal agencies would have needed to use fifteen laws to control all exposures to the substances from its manufacture, use, and disposal. In terms of the hypothetical presented above, at least seven statutory programs under the auspices of EPA might apply to BO and BOC. They are the Toxic Substances Control Act, the Federal Insecticide, Fungicide and Rodenticide Act, Safe Drinking Water Act, Clean Water Act, Clean Air Act, Resource Conservation and Recovery Act, and the Comprehensive Environmental Response, Compensation and Liability Act.

These statutes approach toxic substances from a variety of perspectives and purposes. Some protect human health (for example, the Clean Air Act) and others were designed with the protection of natural resources (for example, the Ocean Dumping Act or Clean Water Act). Some laws were passed to prevent risky environmental exposures (for instance, the Resource Conservation and Recovery Act), and others were enacted to remedy hazardous risks once they occur (for example, the Superfund law). Health-based environmental statutory approaches differ widely between themselves in characterizing risks to toxic exposures and in their regulatory response to questions like "how safe is safe?" The Federal Insecticide, Fungicide and Rodenticide Act, for instance, allows for benefit/risk determinations in controlling how a pesticide with toxic characteristics may be used. The Toxic Substances Control Act also mandates the administrator of EPA to weigh the risks and benefits as well as costs of various regulatory responses provided by the act in choosing the appropriate regulatory response. The Clean Air Act prohibits both of these considerations in setting national primary ambient air quality standards. No one act addresses all the manifestations of toxic exposure from a particular chemical. Consistent and effective regulatory responses based on the existing authority in these laws would be serendipitous.

The acts also vary in terms of the burdens of proof they require EPA to meet before the public health risk of a particular substance can be addressed through regulation. For instance, the Clean Air Act has been interpreted by EPA to mandate consideration of hypersensitive populations in determining an adequate margin of safety for setting national ambient air quality standards to protect the public health, but such groups are not mentioned in the provisions of the pesticides law. Further, administrative procedural requirements and the ability of EPA to sanction noncompliance vary significantly between the acts. Their implementation is also complicated by the existence of separate and federally delegated state programs that address toxic chemicals in the environment and by a complex matrix of federal entitlements on which state and local agencies rely for resources and technical assistance to address hazardous substances.

The most commonly suggested solutions to these legislative constraints are to (1) create a universal environmental statute that encompasses all of the existing laws, (2) amend the existing laws to achieve greater consistencies in the way the agency will address science policy and procedures by piggy-backing one statute's requirements onto another's through such legal mysteries as incorporation by reference, or, (3) institute administrative changes that explicitly account for the interrelationship of cross-media pollutant effects so that alternative statutory goals are considered at the same time. While these are sensible recommendations, all three generic approaches appear to share some common obstacles to their implementation.

The Conservation Foundation, in their description of barriers to legislative reform of environmental statutes, accurately notes that even if one could design a model that encompasses all environmental problems associated with a toxic substance, one must still face the formidable forces of the entrenched interest groups associated with the status quo. These stakeholders include not only the lobbying and public interest groups which are so readily associated with various sides of an environmental issue, but also the congressional committees and subcommittees wielding oversight and budgetary authority over the agencies; law firms specializing in directing others through the maze of federal regulatory programs and statutes; bureaucratic divisions and subdivisions whose existence depends, at least in part, on the disparate nature of the present environmental statutes; and the diverse public and private research institutions that support the various interpretations of these environmental positions and authorities.

In terms of our hypothetical, one could identify at least ten of House and Senate subcommittees (out of seventeen involved in environmental oversight) and equally numerous trade associations[3] and public interest/citizen groups[4] that might be involved in the development of a regulatory action on a particular chemical or family of chemicals. In addition, four program offices at EPA (The Office of Pesticides and Toxic Substances, Office of Water, Office of Solid Waste and Emergency Response, and Office of Air and Radiation) might each become directly involved in regulating BO or BOC. Outside EPA, jurisdiction could be exercised by the Food and Drug Administration and the Department of Agriculture over residues of the pesticide in corn, animal feed, and meat. The Occupational Safety and Health Administration might be investigating evidence of health effects on exposed workers as the basis for establishing an occupational health standard.[5] Supporting research on the substances' properties and attendant risk could be going on in several of EPA's seventeen research facilities presently scattered across ten states, with additional investigations ongoing within the several of the National Institutes of Health and the National Toxicology Program, in addition to private institutions funded with grants from the federal government.

The importance of these stakeholder groups to implementing major or minor statutory reform cannot be understated. The situation today is very different from the late sixties and early seventies when politically powerful coalitions galvanized the environmental movement. The interests of the new stakeholder groups of the eighties were spawned from the current fragmented statutory approach, and in some instances, even created them. They are not likely to yield their existing authority or programmatic responsibilities easily.

From a narrow economic view, the issue is one of efficiency versus equity. Social welfare as a whole may be improved by more integrated statutory approaches to toxic or environmental pollution control, but the benefits and costs of improvement will be borne by different groups; all will not share equally. Centers of authority (power) and resources (wealth) will shift with integrated statutory control, with resultant winners and losers. For example, public interest groups may perceive that valuable gains and concessions made on a piece-by-piece basis will be lost in the new order, or that their energies will be focused on a moving target that has too many bull's-eyes for effective participation. Lobbying organizations may also fear that their raison d'être will evaporate with a more simplified legislative framework and that they will find an integrated scheme too difficult to dissect into manageable pieces. Members of Congress may fear the loss of their committee assignments, constituencies, and even public exposure. Extolling the virtues of more effective environmental protection through toxics integration may be valid, but voting to put your own head on the block may not be politically sound.

One could also argue that the general public also constitutes a stakeholder with a vested interest in environmental management under the current fragmented approach. Although environmental statutes are often written in terms of benefiting the general public, subpopulations or geographic areas are often singled out for particular treatment under the individual laws. For instance, it is clear that areas that experience polluted air are not the same areas that benefit most from the goals of the Clean Water Act. Similarly, residents near abandoned or leaking hazardous waste sites may not be as likely to see or care about the benefits from clean air programs. While this point of view may seriously undervalue the importance that the body politic places on environmental values as a whole, the point to be made here is that if the public organizes itself most effectively to deal with particular environmental issues on regional bases, they might not be so effectively involved if the environmental acts were lumped together.

It is also important to recognize the degree to which public and private industry would be directly and adversely affected by integrated environmental management techniques which would almost certainly demand new kinds of technological solutions that no longer shift wastes from one media to another. These stakeholders in the current approach are easily identifiable and include the pollution control service industry, pollution control manufacturers, as well as the regulated industries which have made considerable investments in environmental controls mandated by goals incorporated in present legislation. The statutory demands for fragmented environmental solutions have also created a fragmented industrial base which produces engineering technology that may not be fully capable of being integrated.

The foregoing is not to suggest that all of the current actors in environmental protection will do their best to block efforts to legislatively integrate the nation's environmental programs. After all, an environmental movement in the eighties is not inconceivable given the political power of a number of environmental factions that could find a uniting theme in toxics integration. There are also several examples, in fact, of toxics activities at the state level that suggest the support of a large number of stakeholders in more effective treatment of toxic substances and hazardous wastes. The Conservation Foundation has noted that Section 28 grants under the Toxic Substances Control Act have enabled nine states to improve the coordination of governmental responsibilities for toxic substances control within their jurisdictions. These include Illinois, New York, Maryland, Michigan, California, and New Jersey. Although these efforts fall short of complete statutory reform by consolidation of all the relevant administrative codes bearing on the environment, they appear to have the support of the various stakeholder groups in the states. Yet efforts at the national level to pass various regulatory reform bills like Senator Bumper's bill to make regulatory programs more consistent suggest results to the contrary. The success of similar efforts (most notably and most recently, reform of the Federal Tax Code) does not bode well for radical change of the way we currently legislate environmental regulation.

Current Research Efforts

Efforts are ongoing in EPA, using the agency's own staff in the Offices of Planning and Policy Evaluation (OPPE) and Office of Enforcement and Compliance Monitoring (OECM), with outside consultants, to study the agency's current statutory authorities for consistency with respect to risk evaluation, economics, enforcement, and civil penalty authority. A division in OPPE called the Integrated Environmental Management Division (IEMD) has attempted to compare and evaluate cross-statutory responsibilities and the basis for regulatory responses as a method for evaluating needs and opportunities for regulatory prioritization in EPA. Some of the administrative reforms suggested by the analysis are discussed in the following section in more detail. EPA's activities in this area are described within the agency as preliminary to proposals for legislative amendments that promote consistency of environmental laws. Members in the enforcement office also suggest that the studies are being evaluated as a basis for a consolidated environmental act.

In addition, a longer term effort to develop proposals for statutory reform has recently been initiated in OPPE. While this project is not looking at an integrated statute per se, it has identified potential statutory issues or problems that cut across all environmental laws. These include, for example, effectiveness of citizen enforcement provisions, and the usefulness and applicability of compliance orders and administrative deadlines.

Significantly, a few private companies such as the Minnesota Mining and Manufacturing Company (3M) perceive a need to create technological "closed loops" for integrating the control of residuals at manufacturing facilities. These efforts are driven in part by the recognition of terrific liability for past waste disposal practices and also by economic recycling and reclamation of residuals back into the production process. Efforts by the authors to identify any other serious efforts among the private sector to compose an integrated environmental approach at this time have yielded no ongoing research, although considerable interest has been expressed by a number of organizations with whom we spoke with in this as a research topic. At least one institution, the Conservation Foundation, has begun work funded by a major private foundation that might eventually lead to such a project. Interestingly, many researchers we spoke with about toxics integration expressed skepticism at the viability of the idea on the basis that managerial and administrative constraints existed under the present system which would prohibit this as a practical solution to toxics integration problems.

Long-term Research Goals

Before an integrated environmental statute can be designed, considerable groundwork must be laid to determine whether such a goal is desirable, or whether consistency in approaching potential hazards to humans and the environment can be introduced more effectively in another manner. The desirability of such a statute cannot be measured by the failure of the existing system alone, but also in conjunction with the political tenability of the idea itself. Is this a justifiable solution or will its fashioning simply delay more effective regulatory responses while new statutory ''kinks'' are ironed out? Are there valid reasons for regulating classes of toxic substances in one or another media differently? Can ''command and control'' environmental statutes be made more consistent with one another or must we start from scratch? What potential does the Toxic Substances Control Act provide for across-the-board regulation of potentially hazardous substances and what are its statutory weaknesses?

The recommendations for technical analysis and research presented in the other chapters in this book must also be accompanied by better efforts to understand the political science ramifications of the policy behind environmental legislation and reform. Specifically, the roles and objective functions of the various stakeholder groups must be more carefully detailed and modeled to discover how consistency in environmental laws could be achieved given stakeholder resistance. It has been suggested that without some sort of compensation scheme or reallocation of power for the losing stakeholder groups, legislative reform efforts may be destined to fail. This appears to require an in-depth understanding of the payoff functions for the vested interests and alternative approaches for utilizing this information to facilitate acceptance of reform initiatives. There is ample history to learn from already, and an assessment of the political factions in the environmental sector today and fifteen years ago might be a very interesting first step. An improved understanding of the success and failure of other legislative reform initiatives might also assist in the design of an effective political strategy for integrating toxic substances control.

ADMINISTRATIVE REFORM

Statement of the Problem

In 1970 when EPA was created by executive order, the agency was designed expressly to fulfill a perceived need for consolidated integration of various environmental responsibilities shared by a number of existing federal agencies. In the ensuing ten years, EPA became responsible for the implementation of a number of environmental laws designed to deal with the exposure and damage of public health and natural resources from a number of human activities, including the manufacture of toxic substances. Nevertheless, programmatic emphases tended more and more to dominate the agency's mission. At the same time, "multimedia offices" such as the Office of Policy and Management, the Office of Research and Development, and the Offices of the General Counsel and Enforcement offered the continued potential for a strong intermedia focus within the agency. There are now twelve program line offices in EPA with counterparts in each of EPA's ten federal regions. In addition, there are staff offices in the agency responsible for agencywide activities including enforcement and compliance, policy, research and development, and administration. Nonetheless, the current EPA management and administrative structure can be best characterized today as a direct response to the single media approach of enabling environmental laws and an amalgam of ad hoc attempts to integrate the control of toxics on a multimedia basis.

Substantive program offices with on-line environmental responsibilities are broken out by media which generally follow statutory lines (for example, the Offices of Air, Water, and Solid Waste). The exception may be the Office of Groundwater, but it presently lacks a clear statutory mandate and is constituted within the Office of Water. Each program office has developed its own "in-house" expertise in engineering, economics, and science. There is little required interaction between the program offices, except for that which is provided in the form of joint-office task forces which have reemerged in the past fifteen months. These interoffice groups appear to supply little interaction, however, on the staff level and offer little incentive to view regulatory initiatives over the long term. There is also a question of which office, using what budget, takes the lead in actions if the need emerges from a task force's consideration of an issue. Avenues for formal integration are thus limited.

The most obvious exception to this media-by-media approach, of course, is the Administrator's Office. In addition, potential avenues for integration exist in the red-border review of major regulatory proposals by all EPA assistant administrators. EPA also houses several program offices that in theory cut across the media-specific functions. The Office of Planning, Policy and Evaluation (OPPE), an assistant administrator-level program in which most of the current administrative reforms are being developed, is the most visible of these. Others include the Office of Research and Development (ORD), the Office of Enforcement and Compliance Monitoring (OECM), and the Office of Administration and Resource Management.

The ability of these institutional structures to implement integration reform activities appears limited, however, by several factors. First, none of the so-called "integrated" program offices enjoy a statutory basis for their existence. While some may draw responsiblities and authorities from various statutes, they do not have the fall-back justification of a legislative mandate for what they may want to promote. In fact, it may be observed that ORD and OPPE are often forced to work as consultants to the program offices, continually selling their regulatory reform agendas like wares in hopes of finding program office support at budget time. The fact that even these "cross-program" offices become segregated into media-specific divisions is testimony to the influence of the other program offices.

Further, the success of initiatives for regulatory integration often hinges on the strength of support for such activities in the administrator's office and the willingness of EPA's leadership to inject the policy office into specific regulatory debates. In most recent times, for example, the ebb and flow of OPPE's strength in overall EPA policy can be clearly matched to the views of the particular administrator. The level of this support has varied considerably over the different EPA administrations with the current administration demonstrating perhaps the highest level of commitment to factoring the duties and analyses of OPPE to day-to-day agency management. Even this support, however, may not be sufficient to induce program offices to adopt new approaches or sustain regulatory reform initiatives.[6]

Second, the current EPA administration structure has developed, like many others, into an entrenched bureaucracy where the beneficiaries of the status quo are less likely to be receptive to reform measures designed to share authorities and resources and to accept the inherent trade-offs of integration. One only has to note the suspicion that accompanies OPPE studies of program office activities and the reticence with which OPPE recommendations are at least initially greeted to see the constraints imposed by the existing bureaucratic power structures.

Finally administrative and managerial reforms, no matter how carefully structured and well-intentioned, will be limited to a greater or lesser degree by the underlying statutory authorities. While some environmental acts and programs explicitly or implicitly require some consideration of cross-media impacts, most do not. For instance, with a few exceptions, present statutory schemes for granting permits are implemented under separate laws and rarely at the same time so that the impact of releases into the environment are rarely evaluated simultaneously. It should also be noted that the chronology of regulation under each of the laws also has had the effect of compounding a source's inability to address later environmental problems that may present more serious hazards. Also, where toxic integration implies use of alternative control techniques, the existing laws still define certain threshold requirements and define how they will be accomplished. For instance, the Clean Water Act still requires technology-based effluent guidelines and the Clean Air Act will still call for health-based national ambient air quality goals. It is true that several of the more recent acts and amendments provide for a wider range of regulatory alternatives (TSCA and the 1978 FIFRA Amendments are the most obvious and CERCLA may be the extreme—its solutions are to be made on a case-by-case basis). Nevertheless, the agency decisions under all these laws may only be stop-gap measures. Taken together, then, the universe of environmental statutes only consider a small set of regulatory techniques, far less than are suggested by most proponents of toxics integration.

The current structure of EPA is not, of course, completely irrational. Faced with environmental problems that often are linked between media, and are sometimes characterized by synergies and multiple risk exposures, effective management of environmental risk, many will argue, still necessitates that a problem be broken down into subtasks and assigned to different sections of the agency. From a management perspective, smaller groups with narrowly defined goals can be more effectively directed, measured, and tracked. Moreover, this type of management structure provides for both internal and external accountability. It is the responsibility of the top layer to integrate and coordinate the control of toxic substances when the distribution and behavior of pollutants create a need for toxics integration. Such a system also facilitates tracking performance, with small enough units to measure individuals' performance. An extreme response to the need for better toxics integration is for the agency to be divided by substantive skills. Work groups or task forces could be constructed for any given substance or regulatory initiative. Such an unwieldy machinery for decision making might build more delay and tension, however, than exists in the present system.

Current Research

By far, most of the current research on toxics integration has focused on administrative and managerial reforms. EPA's current management has inaugurated several administrative reform initiatives to review toxics integration efforts, including the establishment of several external task forces to review related management problems and internal interoffice work groups. It has placed great emphasis, for instance, on the establishment of the cancer assessment, exposure assessment, and reproductive hazard work groups in the development of internal risk assessment guidelines. It has also enhanced the mission of the Integrated Environmental Management Division in OPPE by giving the office greater exposure in and outside the agency. This is being done at a time when IEMD is completing several significant studies of toxics residuals management in the steel and chemical industries, and statistical analyses of monitoring data on sources of toxic air emissions in Philadelphia and Baltimore. In addition, the Office of

Toxics Integration in the Office of Toxic Substances provides coordination of development of toxicological data within the agency.

The IEMD work over the past three years provides a particularly promising area of research for toxics integration. It relies, in part, on statistical and analytic approaches developed in research conducted by Resources for the Future (RFF) and the Office of Toxics Integration. These techniques attempt to measure and to estimate the differences in the quantity and type of residual wastes released by selected industries into the different media using assumptions concerning product characteristics, manufacturing processes, and pollution controls. The IEMD has also developed industry-specific, cross-media pollution abatement models for the iron and steel and paper and pulp industry segments that are designed to estimate the reduction in human health risks attributable to the adoption of various sets of abatement techniques. The division is currently developing a model for the chemical and refinery industry segments. At present, however, these research projects are designed as tools to analyze and evaluate methods of maximizing the agency's use of its regulatory resources not necessarily for toxics integration. These models could also be used to identify those pollutants with cross-media impacts, and they may eventually be used to demonstrate the desirability of more formal integrated approaches. The IEMD has also been assigned the responsibility in EPA for generating analyses of risk assessment requirements and burdens of proof necessary for agency responses. In a case study, it is conducting a study of the sources of toxic substances pollution of groundwater in the Santa Clara Valley in California.

At the request of the Administrator's Office, the Office of Enforcement and Compliance Monitoring is also presently preparing consolidated enforcement guidance under all statutory authorities for regional offices. This document, to be released in September, may also provide a basis for administration amendments to some of the existing acts, including the Clean Water Act which presently does not provide, unlike the other environmental statutes, a basis for levying civil penalties for noncompliance.

In addition to these federal activities, many states, some of whom are receiving Section 28 TSCA grants mentioned above, are studying or attempting to implement techniques for integrating the control of toxic substances. These include methods for better resource management, more accessible and generally available data bases for regulatory decision making, and better ways of facilitating public involvement in decisions regarding toxic substances. For example, New York is using its Environmental Quality Act, which requires permitting authorities to consider all environmental impacts and permitting requirements concurrently, to ensure releases into one environmental medium will not create problems elsewhere. The University of Texas with the assistance of the Conservation Foundation is also studying the consolidation of Texas's environmental permitting requirements.[7] The New Jersey Department of Environmental Protection is attempting to integrate its pollution inventory by surveying industries in the state about production inputs and the volume of environmental releases, and Michigan's Department of Natural Resources is creating a computer bank of production volume data and toxicity of chemicals used or produced in the state. In Maryland, Governor Hughes has appointed a task force to oversee important toxic problems in the state. Several years ago, Maryland also integrated the primary regulatory authority for the environment in the state's Department of Health and Mental Hygiene, so that public health and medical resources would be available to the agency on a day-to-day working basis. Such investigations and experience with integrated permitting and data bases at the state level will aid federal efforts at coordinate information and regulatory activities on toxic substances and hazardous wastes.

Long-term Research

Further policy research is needed to track the efforts of present institutions that are implementing new administrative and managerial techniques for integrating information about toxic substances and prioritizing regulatory actions on this basis. Institutional reorganization within EPA in response to specific environmental problems may also provide the basis for long-term tracking of other looming environmental problems. For instance, case studies of the development of the Office of Groundwater and the evolution of the Groundwater Strategy or of EPA's current acid rain activities could provide valuable information about administrative constraints to toxics integration efforts since each of these areas involve toxic problems with multimedia dimensions.

A great deal of regulatory reform/relief/alternative controls research has been funded by the federal government and private institutions over the last six years. One prominent research topic has been the use of economic incentives such as effluent fees and taxes to achieve more efficient pollution reduction. The emissions trading policy currently being finalized by EPA is one area in which this research has resulted in regulatory policy. The Steel Compliance Extension Act that supplied incentives for capitalizing the depressed steel industry by rolling back compliance deadlines in the Clean Air Act is an example of how such studies led to the amendment of existing laws.

However, little follow-up has been done to collect empirical information about the way institutions act in response to these incentives. In order to facilitate the types of institutional changes that will be necessary to accommodate toxics integration techniques, we need more information about how private and public institutions respond to regulatory reform initiatives. For example, EPA as well as private firms may try to maximize internal goals other than economic efficiency. Some large corporations such as the 3M Company have already begun comprehensive integration of their environmental programs and approaches in order to reduce hazardous wastes and maximize pollution abatement across plants. The identification of other companies like 3M and studies of their organization structures could yield valuable insights on company environmental management and decision making. These in turn may provide ideas for structuring or reforming the management of environmental regulatory programs. Case studies of environmental auditing may lay the basis for some of this longer term research. Another related area for research is the development of models and incentives for producing pollution abatement technologies that reduce total waste loadings, instead of shifting waste to another media.

SCIENCE POLICY

Statement of the Problem

Many of the areas in which we simply require more scientific information about the mechanisms and dynamics of cross-media pollution effects have been addressed by other members of this panel. For example, the basis for identifying whether there is a subset of toxic substances that have these intermedia effects or whether all chemicals to a greater or lesser degree enjoy similar properties is not clear. Similarly, the mechanisms of chemical transformation under aerobic and anaerobic conditions in different environmental media is not understood. The mechanisms of pollution interface between different media requires much more work be done in mass balances for particular pollutants. These types of questions are primarily technical, scientific ones and we turn our attention instead to an area that we call "science policy."

In the following discussion, the term "science policy" is used as a shorthand for the way that decision makers develop information about cross-media pollution problems and apply it in environmental policy. This would encompass, for example, the way decision makers evaluate the risk from toxic substances, but not how scientists estimate these risks. Our thesis is that science policy drives the development of science in certain directions toward the solution of environmental problems. In this sense, science policy may have, in some instances, failed to generate the type of scientific information that addresses cross-media pollutant exposures or their intermedia transfer. A corollary question seems to be what factors would encourage the development of the scientific underpinnings necessary to integrate the regulation and control of hazardous environmental substances? Thus by itself, it is not enough to identify the specific gaps in our scientific knowledge about these problems. The political, legal, and institutional arrangements that create the demand for this type of information and data must also be considered. In addition, we must ask ourselves whether legal and administrative constraints alone inhibit toxics integration and its scientific support. Finally, we should ask if there is something about the discipline of science itself, or the way that scientists and engineers are trained that may inhibit discussions of cross-media pollution effects and controls.

Legislation is clearly one impetus to the development of science in the environmental field that has lent itself to engineering solutions for pollution control. Current statutory requirements create the demand for certain kinds of technological solutions which, in turn, drive scientific investigations toward the development of specific data and measurement techniques. This was clearly the case in the almost unbelievably swift evolution of catalytic converter technology at the turn of the last decade. A development generated, in large part, by the 90 percent reductions in VOC and NOx exhaust emissions mandated by the 1970 amendments to the Clean Air Act. Legislation has also led to the development of certain types of technology, rather than others. The effluent guidelines and NPDES permit system promulgated to implement the Clean Water Act demanded the development of end of pipe pollution abatement technology and sampling techniques for measuring pollutants at the pollution outfalls (end of pipes). Unfortunately, many of the technological solutions that legislation developed the need for did not consider the interrelationship of the media into which wastes from the technology were disposed. The scrubber technologies developed to meet the New Source Performance Standards under the Clean Air Act are the classic example. These technologies may have in fact exacerbated cross-media pollution problems by the creation of vast amounts of toxic sludges that needed to be disposed, and were generally buried on land, with resulting groundwater, and sometimes, surface water contamination.

There are two statutes of more recent vintage than the media-type environmental control laws whose legislative history reveals the perceived need to permit broader administrative flexibility in terms of weighing cross environmental and human health impacts. The existence of these statutes would tend to diffuse arguments that statutes are the primary constraint to science policy for toxics integration. As we mentioned above, the 1976 Toxic Substances Control Act and 1972, 1975, and 1978 amendments to the Federal Insecticide, Fungicide, and Rodenticide Act allow for examinations of chemical exposure across media, and across geographically selected areas. TSCA also may require manufacturers, as a condition for the premarket approval of a substance, to consider synergistic relationships with other chemicals. Under these laws, the administrator may require, as a condition for registration of a substance, that manufacturers of particular chemicals or classes of chemicals develop data about cross-media exposures and environmental fate and transport. He may also insist on premanufacture testing that requires potential synergistic relationships between chemicals to be considered. For instance, in the case of our hypothetical, the administrator could require any possible synergies between Lesritts and Rower be investigated by the producers of the compounds through the promulgation of a testing rule for existing chemicals, or through a request for data under an existing pesticide registration. The kind of broad investigatory authority in TSCA and FIFRA was in fact used by a task force to investigate routes of human exposure to the compound ethyl dibromide (EDB). This study led in turn to the discovery of EDB in numerous groundwater supplies throughout the country. TSCA and FIFRA also allow for a broad range of regulatory responses ranging from total bans on the production and use of chemicals that pose unreasonable human or environmental risks, to alternative work practices, labeling requirements, and even public education advisory circulars about their use and disposal.

Still, it is likely that the older acts have driven and continue to drive the need for certain scientific information and engineering solutions which continue to ignore many of the cross-media implications of pollution. As important as these are, however, statutory constraints do not appear to be the sole limitation to the development of toxic integration science. Other impediments to the development and implementation of techniques for integrated sampling, monitoring, modeling, and control can and should be identified.

Administrative Policies and Related Research Needs

The present bureaucratic structure and budgetary process of EPA almost certainly affects the ability of science policy to account for integrated toxic substance impacts. The science needed to support cross-media exposure assessments and control responses will not evolve without bureaucratic or programmatic mechanisms for allocating federal research dollars toward this goal. For example, the existence of a centralized Office of Research and Development at EPA headquarters may fail to produce integrated science policy or the requisite scientific tools for risk assessment and data sampling even though it does have a multimedia perspective. This is because ORD, like OPPE, must be responsive to the program offices and the budgetary process through which its research resources are "divided" up. Long-term and short-term research goals within and outside of the agency are structured to develop scientific information to support the ongoing bureaucratic functions of EPA. Without management integration or the articulation of toxics integration policy from the Administrator's Office or possibly Congress, the impetus for the development of the necessary science will not exist, and even then, may require reallocation of other offices' budgets.

There are presently several serious administrative initiatives to integrate science policy on the control of toxic substances over various media. Like the Integrated Environmental Management Division in OPPE which has produced internal agency management guidance on doing risk assessments to guide science policy at EPA, three "science policy" risk assessment groups within ORD have been charged with the development of risk assessment guidelines for cancer, reproductive hazards, exposure, mutagenicity, and developmental effects. These are in various stages of completion with several expected to go to the Administrator's Office for publication as notices in the Federal Register in October. In both of these projects, the ORD and OPPE offices report that one common obstacle to the use and development of risk assessment tools, be they scientific or management oriented, is that there is no central data repository for health or environmental exposure data in EPA and programs do not customarily share data received pursuant to statutory mandates.[8] Even if such a repository existed, data might be virtually impossible to compare because sampling, measurement, and extrapolation techniques differ widely across EPA programs and indeed, between agencies. Within EPA, for example, different program offices have instituted the use of different testing protocols for applicants to meet data requirements under different acts. These problems are shared by and with other federal agencies such as the Departments of Labor and Health and Human Services which also collect toxicity data on specific chemicals.

To address some of the manifestations of these data problems, the National Toxicology Program and its member agencies are attempting to coordinate research on toxic chemicals and to develop new and better testing protocols for producing the types of information these regulatory agencies require. In addition, the old Interagency Regulatory Liaison Group has been recently reconstituted as an interagency risk management group to discuss interagency risk assessment guidelines similar to the ones currently under development at EPA. The White House's Office of Science and Policy also released a set of cancer principles several months ago in an effort to provide assistance across agencies in cancer assessments. This latter document has been both praised and criticized for its conservatism. Prior cancer policies promulgated by a handful of agencies including EPA and the Occupational Safety and Health Administration during the 1970s have been similarly attacked.

These administrative attempts to develop guidelines and cancer policies are a science policy impetus to standardize the scientific community's state of knowledge about cancer etiology and risk assessment techniques in order to fashion consistent regulatory responses. The problem regulators say, and some scientists agree, is that there is significant disagreement over many definitional problems and methodologies in the scientific community itself which constrain the development of science policy strategies such as these. The coordination of the extreme positions often necessary in molding consensus science policy like the cancer documents often results in assertions that are so broad that the document becomes essentially useless scientifically and administratively. This is an important topic that should be evaluated in the context of intermedia environmental pollution concerns.

Another related regulatory current, also with a long and uneven history, is the development of risk assessment policies by government. Under the second Ruckleshaus administration, for instance, EPA added another chapter by attempting to separate the assessment of health risks from the policy of managing risk. It is possible that if this agenda, which calls for isolating science from science policy, is taken to its extreme, mechanisms for ensuring the kinds of science information needed for toxic integration will not emerge. In addition, without strong agency leadership, this new science policy may not guarantee any better than prior risk assessment policies the development of integrated data bases about how compounds are emitted from myriad sources in the environment, are transported, and transformed across media. One institutional solution may be in using a group like the Science Advisory Board, which presently addresses technical needs for assessing data in EPA, to act as a bridge between risk assessors and risk managers in recommending science policy for integrating toxics control. Without strong coordination of science policy needs, the present vogue to separate the science from the management of risk may further compound efforts to develop the integrated information needed by limiting existing communication channels between scientists and policy decision makers.

Scientific Disciplines and Research Needs

It is difficult for the authors, given their respective training, to competently address the question whether there are constraints that may be present in the scientific disciplines themselves that impede the development of toxic integration strategies. Is there, for instance, something in the way scientists view themselves or in the way they are trained that perhaps narrows their focus in investigations to isolate cause-and-effect mechanisms? As we were discussing this idea, we recalled a criticism cited routinely in current legal and business literature that students are now so narrowly taught the mechanics of addressing small subsets of problems that they rarely emerge from universities with comprehensive views about how these problem-solving techniques merge into broad practical skills. Educators frequently defend this training as providing the skills that businesses and law firms require for their specialized fields.[9]

Other Long-term Research Goals

The various studies presented here have identified a number of areas where the sciences can generate information about the way chemicals disperse throughout the environment and models for evaluating their cumulative risks and control. We know, however, of no efforts to study the institutionalization of science goals into what we have roughly equated as "science policy." For instance, why is it that the cancer policy initiatives have consistently failed to produce science policy? A long-term evaluation of the policy goals for coordinating these efforts needs to be conducted to ensure that research goals can be directed toward effective and efficient environmental solutions.

NOTES

1. These are the Clean Air Act; Clean Water Act; Safe Drinking Water Act; Federal Insecticide, Fungicide and Rodenticide Act; Resource Conservation and Recovery Act; Comprehensive Environmental Response, Compensation and Liability Act (Superfund); Noise Act; Marine Protection, Research and Sanctuaries Act (Ocean Dumping Act); Toxic Substances Control Act; the Environmental Research Development and Demonstration Act; and the Uranium Mill Tailings Radiation Control Act. In some instances, the National Environmental Policy Act, administered by the Council on Environmental Quality, or its functional equivalent must also be implemented by EPA.
2. The most obvious are the Department of Interior, Department of Health and Human Serv-

ices, Department of Labor, Department of Agriculture, Department of Energy, Nuclear Regulatory Commission, Consumer Product and Safety Commission, Department of Commerce, and Department of Transportation. The State Department also maintains an office of environmental affairs.

3. For instance the Chemical Manufacturers Association, Specialty Chemicals Manufacturing Association, National Agricultural Chemicals Association, National Governor's Association, Adhesives Union, American Petroleum Institute, and the National Refinery Association.

4. For example, the Natural Resources Defense Council, the AFL/CIO, the Coalition for Reproductive Hazards for Workers, Migratory Farmworkers League, and Citizens for a Better Environment.

5. Note that EPA, and OSHA, would set farm worker protection work practices and applicator certification standards if residues posed unreasonable risks to human health.

6. For example, the regulatory impact requirements of E.O. 12291 which calls for the examination of regulatory alternatives and a description of current limiting statutory authority for implementing any of these, although coming from the highest administrative level in the federal government, have had a mixed success in terms of the balance intended by their application.

7. It should be remembered that in 1980, EPA failed in its attempt to consolidate several of its permitting programs under RCRA, CWA, CAA, and SDWA. An analysis of this result and the litigation over the issue would provide valuable insights for future attempts to integrate different program's permitting requirements.

8. These offices continue to rely heavily, it appears, on the Chemical Information System (CIS) and the Chemical Substances Information Network (CSIN) except where specific substances have been referred to them by a media program office.

9. Daniel Yankelovich in the first issue of *Issues in Science and Technology* has suggested that scientists and scientific institutions must "rewrite" their social contract and play a more significant role in public policy matters to sustain science and technological society. *See* D. Yankelovich, "Science and Public Policy," *Issues in Science and Technology* (National Academy Press: Fall 1984).

REFERENCES

Alvin L. Alm, "Building a Better EPA," *EPA Journal,* p. 16 (June 1984).

Conservation Foundation, "Controlling Cross-Media Pollutants," *State of the Environment: An Assessment at Mid-Decade* (1984).

William Eichbaum, "Managing Hazardous and Toxic Materials in Maryland: Key Concepts in Integration," 2 *Environmental Forum,* p. 13 (December 1983).

"Misgivings Over a 'Consensus' on Carcinogenesis," *Chemical Week,* pp. 10–12 (August 29, 1984).

National Academy of Public Administration, *Steps Toward a Stable Future: An Assessment of the Budget and Personnel Processes of the Environmental Protection Agency* (May 1984).

Office of Science and Technology Policy, *Chemical Carcinogens: Notice of Review of Science and Its Associated Principles* 49 Fed.Reg. 21594 (May 22, 1984).

U.S. Department of Health and Human Services, *National Toxicology Program Review of Current DHS, DOE and EPA Research Related to Toxicology: FY 1984,* NTP–84–024 (February 1984).

U.S. Environmental Protection Agency, *CASR Toxics Integration Series* (2 Vols.), EPA 560/THS–84–001a (February 1984).

U.S. Environmental Protection Agency, *Federal Activities in Toxic Substances,* EPA 560/THS–83–007 (September 1983).

U.S. Environmental Protection Agency, *Final Report of the Toxics Integration Task Force, Risk Management Subcommittee,* Memo to Deputy Administrator Alvin L. Alm (October 28, 1983).

U.S. Environmental Protection Agency, First Report of the Interagency Toxic Substances Data Committee, EPA 500/1–82–003 (November 1980).

U.S. Environmental Protection Agency, *Program Description of Integrated Environmental Management Division in Office of Policy Analysis, OPPE* (undated).

U.S. General Accounting Office, *EPA's Efforts to Identify and Control Harmful Chemicals in Use,* GAO/RCED–84–100 (June 13, 1984).

8
MODELING OF POLLUTANT TRANSPORT AND ACCUMULATION IN A MULTIMEDIA ENVIRONMENT

Yoram Cohen

INTRODUCTION

Pollutants released to the environment are distributed among the different environmental media (i.e., atmosphere, soil, water, biota) as the result of complex physical, chemical, and biological processes. The hazardous potential of chemical pollutants in the environment depends upon the degree of multimedia exposure of human and ecological receptors to these chemicals. Therefore, a multimedia understanding of pollutant behavior in the environment is essential for the early assessment of the associated environmental impact (Freed et al., 1977). This is of particular concern for chemicals that are toxic, persistent, and subject to accumulation in the environment.

Due to the fragmentary nature of environmental laws, governmental agencies are faced with a formidable task of trying to follow policies that are based on conclusions derived from single-medium programs (Conway, 1982; ref. 20, 37). Single-medium analyses do not reflect the interrelationships or interactions among air, water, and terrestial environments. This is also evident in the existence of separate governmental programs that deal only with single-medium issues. Such individual-medium programs often result in a duplication of effort without achieving a better understanding of pollution as a multimedia problem. Decision makers must recognize that regulations must be formulated based upon a multimedia policy rather than upon the recommendations of disparate single-medium programs (ref. 20).

The large number of current and future pollutants precludes a detailed experimental evaluation of each chemical as to its potential impact on the environment. Experimental investigations of the fate and transport of pollutants *in the environment* can only be conducted after their release to the environment. Moreover, laboratory studies cannot duplicate the dynamics of the natural environment. Although field studies with tracer chemicals are useful, the behavior of tracer elements in the environment may be entirely different from that of the pollutants of concern. In contrast, mathematical models of pollutant fate and transport in a multimedia environment are extremely attractive, for they offer a relatively rapid and inexpensive assessment of potential hazards in the early stages of chemical production or process development. Although there are many practical uses for multimedia models of pollutant transport and accumulation, the ultimate goal of such models is to predict the exposure of a target human population or of ecological receptors to toxic chemicals. Subsequently, it must be determined whether or not the resulting exposure levels will lead to adverse health effects.

The structure of multimedia models and their input and output data may vary depending on the properties of the chemical, spatial, and temporal scales of interest, the release scenario, the lifetime of the chemical, and environmental transport rates. Short-lived chemicals that are rapidly transformed to other chemicals, or transferred to media bordering the region where the emission has occurred, may require the use of local scales, and hence models that can describe short-term fluctuations in the sources and concentrations. Long-lived or slowly transported chemicals require modeling of long range transport and accumulation over long time scales. Depending on the source characteristics and the environmental transport and transformation processes, it may be necessary to predict either long-term average exposure or short-term fluctuations in exposure of specific sensitive receptors. In each case the model has to be adapted to mimic the characteristics of the problem.

It is unlikely that a single comprehensive model could be developed in the near future to predict the transport, transformations, and accumulation of every toxic chemical that has been introduced or will be introduced into the environment (ref. 143). Therefore, the analyst is faced with the need to link together single-medium models, or to formulate specific multimedia models, depending on the required application. In order to maximize the level of useful quantitative predictions, the modeler must carefully consider the required temporal and spatial scale, the complexity of the model in relation to available model input parameters, model interpretation, and validation by laboratory and field measurements. Irrespective of the mathematical complexity of the model, a true forecasting capability can only be achieved through an accurate description of the physical, chemical, and biological intermedia and transformation processes (ref. 119). Obviously any multimedia program, whether theoretical or experimental, must first consider the role of each environmental compartment in an overall multimedia scheme.

It is beyond the scope of this chapter to provide a comprehensive review of all the separate environmental compartments and their incorporation in multimedia modeling schemes. Consequently, this chapter only presents an overview of various topics that are pertinent to multimedia fate and transport modeling efforts. Furthermore, the chapter is restricted to organic compounds which comprise the majority of known toxic pollutants in the environment.

This chapter is divided into three major parts. The first part discusses multimedia modeling of the fate and transport of pollutants in the environment. A specific example of multimedia-compartmental modeling scheme is outlined in order to illustrate the required multidisciplinary nature of multimedia research. Subsequently, a discussion is presented of each of the major components that make up multimedia models. This includes pollution sources, the atmospheric, terrestrial, and aquatic environments. The discussion also encompasses the mutual interactions of the compartments as well as the need for future research. Additionally, the role of microcosms in multimedia research is briefly discussed, and the national problem of hazardous waste disposal sites is presented as an unique example of a multimedia pollution problem. Finally, the conclusions of this chapter are given emphasizing the urgent need for an organized research effort of environmental pollution as a multimedia problem.

MULTIMEDIA MODELING

Multimedia models can be formulated at different levels of detail with regard to both the model structure and the various physical and chemical processes (Swann and Eshenroeder, 1983). These models can be formulated as a collection of well-mixed compartments each representing a different sector of the environment, or they may involve detail modeling of temporal and spatial variations. In the alternative they may be a combination of both. When a site-specific predictive capability is needed spatial models are necessary. Such models require the solution of complex four dimensional space-time partial differential equations. In contrast, the solution of the well-mixed compartmental models requires only the solution of time dependent ordinary differential equations.

Examples of spatial models are the UTM-TOX model (Patterson et al., 1982; Browman et al., 1982) and the ALWAS model (Tucker et al., 1982). These models treat the transport of pollutants in multimedia systems by linking complex simulators of atmospheric, hydrologic, and sediment transport. In contrast, compartmental models such as the Fugacity type models (Mackay et al., 1982), the ADL models (Lyman, 1982) and the MCM model (Cohen and Ryan, 1984) utilize well-mixed compartments.

Regardless of the level of modeling detail and complexity, the results from multimedia models should be carefully studied. When a model predicts that hazardous exposure concentrations are likely to occur, it may be desirable to conduct a sensitivity study to determine whether the predicted exposure levels are sensitive functions of the model parameters. Subsequently, the level of confidence in the model parameters should be reexamined in order to determine the confidence level in the decision-making process based on the model results.

The confidence in the model results will rely heavily on the accuracy of the input variables. Clearly, if the input variable such as the transport and partition parameters are based on a sound understanding of the underlying physical principles, a greater confidence in the model prediction will result. It is emphasized that the model should be tested at all stages of development against available field data. This means that the model development may be an iterative process. Questions that arise should then be answered through further field measurements or further theoretical work designed to formulate and predict the required model parameters.

To summarize, the ultimate goal of environmental multimedia modeling should be to develop mathematical models that are capable of reliable prediction of the transport and accumulation of toxic chemicals in the environment. Model formulation should be based upon measurable or predicted physicochemical properties of the pollutant, the major compartmental components of the environment, and a reasonable knowledge of the dynamics and interactions of transport and transformation reactions.

In recent years considerable effort has been devoted to quantifying the dynamics of pollutant partitioning in the environment (Thibodeaux, 1979; Friedlander and Pruppacher, 1981). Much research, however, remains to be done in order to identify the various mechanisms of intermedia transport, chemical transformations, and sources. In order to illustrate the manner in which the model parameters fit into an overall multimedia scheme, the simple multimedia-compartmental model (Cohen and Ryan, 1984) is discussed in the following section. The model is by no means the most general one, but it will suffice to demonstrate the areas of future concern.

MULTIMEDIA-COMPARTMENTAL MODELS

General Considerations

Although pollutant transport models exist for individual environmental media (i.e., air, water, and soil) the interactions among such media are still poorly understood. In principle, detailed single-medium models can be coupled to yield a multimedia description of environmental transport. Such an approach, however, requires a large number of model parameters that are often unavailable. Moreover, since complex models are difficult to apply and interpret, their regulatory role may be limited.

At the present state of the art of multimedia transport research, the most practical approach is to formulate models that make use of compartmental systems to describe the events occurring in various parts of the environment (see Figure 8–1). These compartments may correspond to actual environmental media or to abstract mathematical constructs designed for computational and modeling convenience. The direct results from such multimedia-compartmental models (MCM models) are in the form of concentration-time profiles for each of the environmental compartments under consideration. Although the spatial description in some of the compartment is sacrificed, quantitative estimates can be obtained regarding the macro-behavior of the environmental system. This information can be utilized to determine pollutant pathways, fluxes and accumulation, environmental "hot spots", and to assess the level of human exposure.

The concept of compartmental modeling is not new in environmental research (Gillett et al., 1974). Compartmental models have been used in modeling the transport of carbon dioxide (Oeschger et al., 1975; Bolin, 1983; Tans, 1980) and trace pollutants in the oceans and across the air-ocean interface (Slinn et al., 1978), in the design and interpretation of multimedia monitoring data (Behar et al., 1979; Wiersma, 1979; Burton, 1981; ref. 117), and in pollutant mass balance studies in the environment (Hantzicker et al., 1975; Neely, 1977a, 1977b; Singh et al., 1979; Wiersma, 1979). In recent years there has also been a growing awareness that MCM models may also aid in comprehensive pollution control strategies (Addison et al., 1983; Hushon et al., 1983; Hedden and Mulkey 1981; Chawgla and Varma, 1982; Neely, 1980).

Many of the early compartmental models were restricted to the determination of steady state mass budgets or concentrations (Cohen, 1981; Cohen and Ryan, 1984). The steady state approach has proven useful in identifying the presence of potential sinks and sources in specific environmental compartments. The drawback of the steady state approach is the loss of information regarding temporal concentration changes. Consequently, exposure variations due to variable sources and climatic conditions cannot be predicted. This deficiency has been recognized in the literature as is evident by the increasing number of dynamic single-medium and multimedia-compartmental models (Bonazountas and Wagner, 1981; Burns et al., 1981; Mackay et al., 1982; Cohen and Ryan, 1984; Neely 1980).

It is emphasized that a properly formulated MCM model should be a true predictive model based on a fundamental description of the governing intermedia transport processes. Many of the existing MCM models have failed to recognize this point and have simply employed transport coefficients which are essentially long time averages of measured flux/concentration ratios or estimates based on first order kinetics without thermodynamic constraints. As a result, these models are not truly predictive, but rather are empirical models that must rely on the adjustment of model parameters using field data.

Modeling Scheme

The global scheme for multimedia-compartmental (MCM) modeling is shown schematically in Figure 8–1. The input variables include media properties, phys-

Figure 8–1. Multimedia-Compartmental Modeling

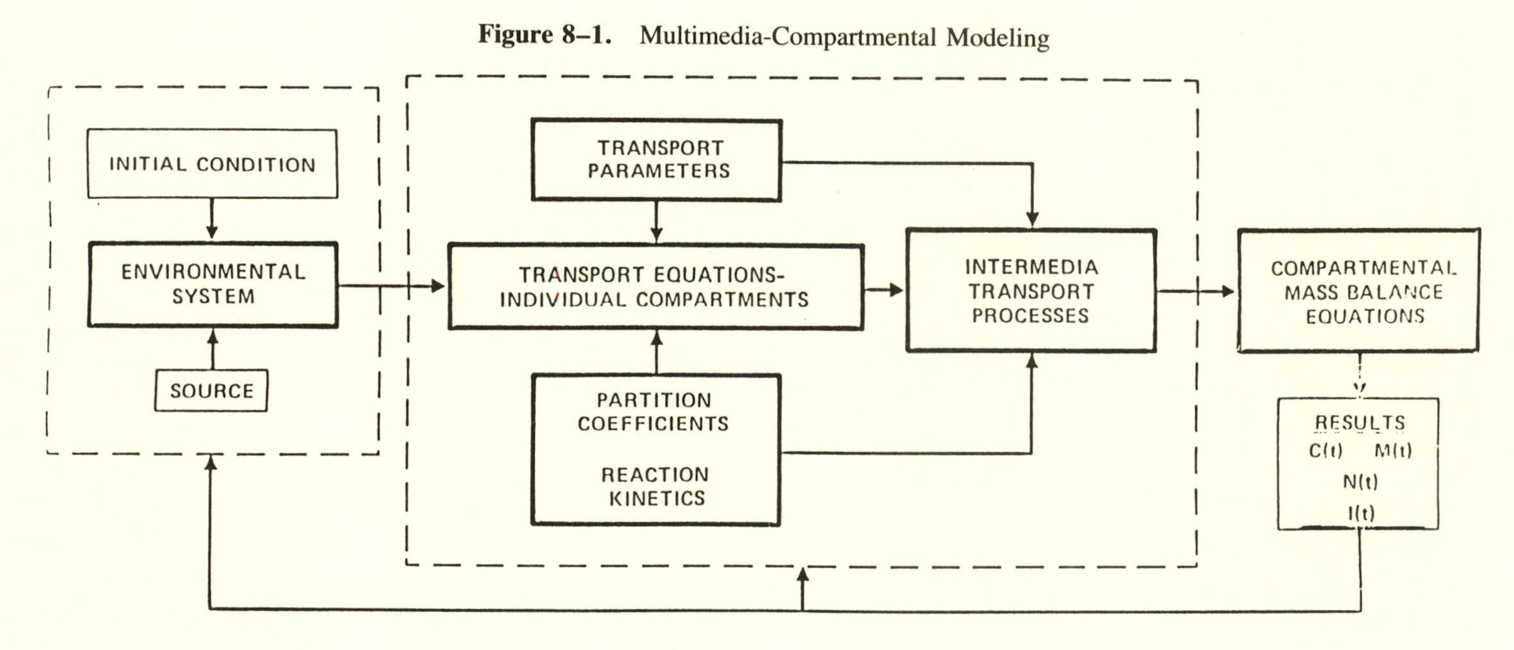

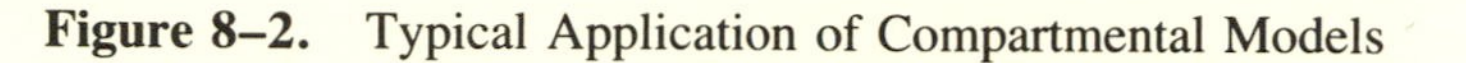

Figure 8–2. Typical Application of Compartmental Models

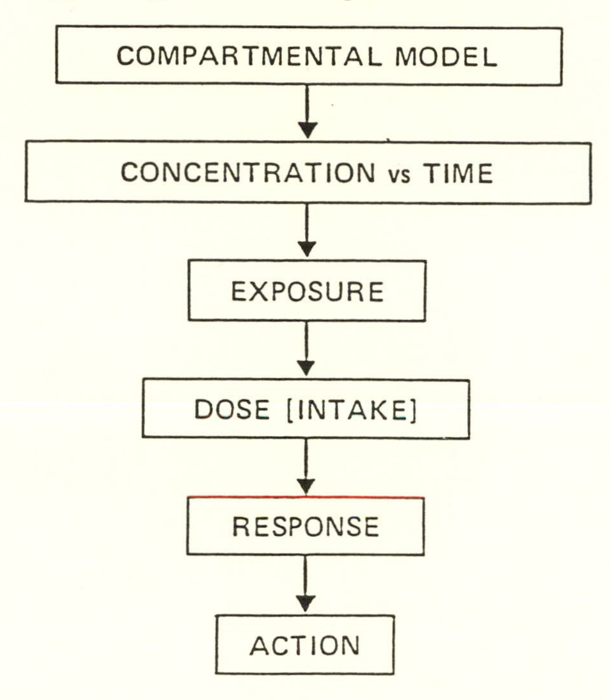

icochemical and thermodynamic properties of the pollutant, partition coefficients, sources, climatic conditions, and initial background concentrations. The output from the model is in the form of concentration versus time profiles for each of the environmental compartments. The resulting time concentration profiles can be coupled with other models designed to determine exposure and hence risk-assessment (see Figure 8–2).

At the present time there is no systematic procedure for constructing a multimedia-compartmental system. The construction of such a system depends upon the specific region under consideration and the judgment of the modeler. For example, the compartmental system that is shown in Figure 8–3 consists of three macrocompartments (atmosphere, water, and land), these are further divided into a total of eight subcompartments, five of which are subdivided into a total of seventeen microcompartments, leading to a total of twenty compartments. Various plausible transport processes in the multimedia environment system are listed in Table 8–1. Although the above compartmental system is certainly not the most general one, it clearly demonstrates that even in a relatively simple compartmental system a great deal of information is required to describe pollutant transport across compartmental boundaries, as well as various chemical transformations within the compartments themselves.

The transport of a given pollutant in a system of N environmental com-

Figure 8–3. An Example of a Multimedia-Compartmental System

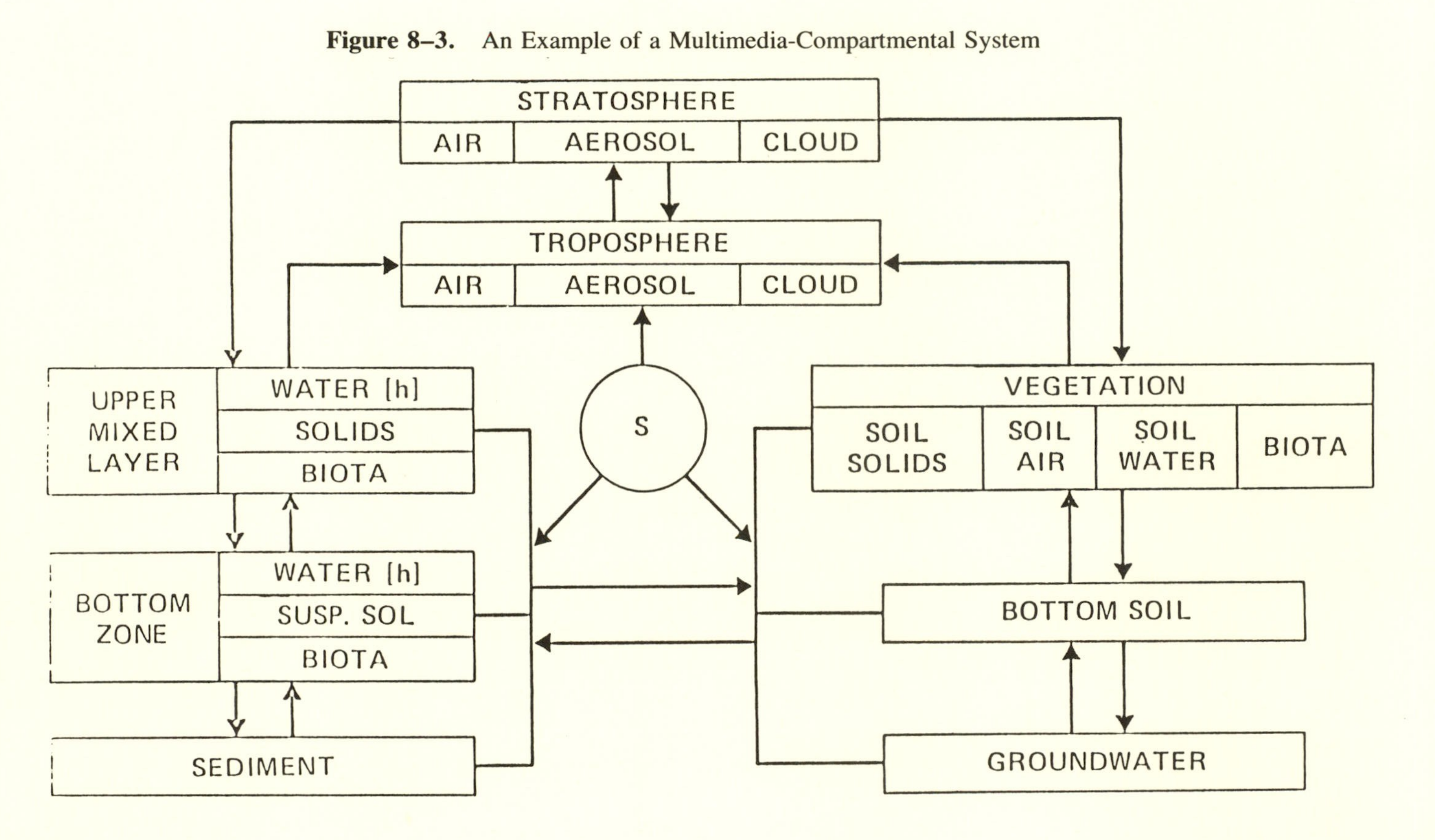

Table 8–1 Summary of Major Transport and Transformation Mechanisms

Atmosphere

Transport from the atmosphere to land and water
- a. Dry deposition of particulate and gaseous pollutants
- b. Precipitation scavenging of gases and aerosols
- c. Adsorption onto particulate matter and subsequent dry and wet deposition

Transport within the atmosphere
- a. Turbulent mixing and convection within the troposphere
- b. Diffusion to stratosphere

Atmospheric transformations
- a. Photochemical degradation by direct absorption of light, or by accepting energy from an excited donor molecules, or by reacting with another chemical that has reached an excited state.
- b. Oxidation by ozone
- c. Reaction with free radicals
- d. Reactions with other chemicals

Water

Transport from water to atmosphere, sediment and organisms
- a. volatilization
- b. sorption by sediment and suspended solids
- c. sedimentation and resuspension of solids
- d. aerosol formation at the air/water interface
- e. uptake and release by biota

Transport within water bodies
- a. Turbulent dispersion and convection due to currents, with shear, and waves in the upper mixed layer
- b. Turbulent mixing and dispersion in confined flows (i.e., mixing and dispersion in rivers)
- c. Diffusion between the upper mixed layer and the bottom layer

Transformations
- a. Biodegradation
- b. Photochemical transformations
- c. Chemical degradation (affected by chemical agents), e.g., hydrolysis, free radical oxidation

Soil

Transport from soil to water, sediment, atmosphere or biota
- a. Dissolution in rain water
- b. Adsorption on soil particles and transport by runoff or wind erosion
- c. Volatilization from soil and vegetation
- d. Leaching into groundwater
- e. resuspension of contaminated soil particles by wind
- f. Uptake by microorganisms, plants, and animals

Transformations
- a. biodegradation
- b. photodegradation at plant and soil surfaces

Source: Compiled by the author.

partments can be described by a set of unsteady mass balance equations. It is assumed that the distribution of chemical species within each compartment is uniform. When such an approximation is invalid, it may be possible to subdivide the compartment into microcompartments in order to approach a condition of a uniform pollutant concentration within each microcompartment. The model equations are expressed by the following set of balance equations:

$$V_i \frac{dC_i}{dt} = \sum_{j=1}^{N} K_{ij} a_{ij} (C^*_{ij} - C_i) + V_i K_i \xi_i C_i + \sum_{j=1}^{N} Q_{ji} C_j - \sum_{j=1}^{N} Q_{ij} C_i + s_i, \qquad i = 1 \ldots N, \qquad 1 \neq j \tag{8.1}$$

The initial concentrations are set to C_i (O) at t=o, and the total source strength (mol/hr) is designated by s_i. C_i is the concentration (mol/cm^3) of the species of interest in compartment i. K_{ij} are the overall mass transfer coefficients in units of m/hr, based on compartment i, for the exchange of mass between compartments and j, a_{ij} is the corresponding interfacial area (m^2) and V_i is the compartmental volume (m^3). The variable C^*as_{ij} is defined as the concentration of the pollutant in compartment in equilibrium with compartment j. The equilibrium relationship is assumed to have the following form:

$$C^*_{ij} = C_j H_{ji} \tag{8.2}$$

in which H_{ji} is the dimensionless i to j partition coefficient. It is noted that $(C^*as_{ij} - C_i)$ is the concentration driving force for internal mass transport constrained by equilibrium.

For simplicity it may be assumed that the pollutant undergoes a first order transformation reaction (chemical or biochemical). If more complex rate expressions are applicable, they can be included without a loss of generality. The reaction rate constant is k_i and the corresponding coefficient ξ_i equals -1 for a degradation reaction, and $+1$ for a production reaction.

The terms $Q_{ji}C_j$ and $Q_{ij}C_i$ represent the pollutant mass flow rates (mol/hr) from compartments j to i, and i to j, respectively, where Q_{ji} and Q_{ij} are the corresponding flow rates (m^3/hr). For example, in a flowing water body such as a river, the water flow into and out from the water compartment is readily identified with the average river flow rate. The advection term for the atmospheric compartment however, must account for the extensive degree of air recirculation. This can be accomplished by considering the atmospheric convective residence time that can be estimated from various atmospheric models. Nonideal flow patterns in the air compartment may alternatively be modeled by subdividing the air compartment into subcompartments with properly defined transport parameters between adjacent compartments. The division of a given environmental compartment into subcompartments and the evaluation of diffusive or convective transport among subcompartments should be guided by information obtained from detailed single-medium models, or from field characterization of mixing characteristics within the compartments. With the above approach it should be possible to relax the assumption of uniform compartments without increasing the computational complexity of multimedia models. As an example, the EXAM model (Burns et al., 1981) and the SEASOIL model (Bonazountas and Wagner, 1981) utilize sub-compartments to account for the nonuniform mixing condition in the aquatic and soil environments, respectively. Although the use of uniform compartments is attractive, it may lead to excessive model parameterization and hence to a loss of a priori predictive capability. Particularly when molecular diffusive transport is the dominant transport mechanism in a given compartment, a spatial transport model for this compartment should be used. For example, the recent MCM model developed at the National Center for Intermedia Transport Research (Cohen and Ryan, 1984) makes use of both subcompartments, knowledge of mixing characteristics, and single-medium models.

Precipitation scavenging of gaseous and particulate pollutants can also be quantified by the advection terms with an appropriate definition of the flow variable Q_{ij}. For example, the concentrations of a given pollutant, C_d and C_p, in the gaseous and particulate form, respectively, in rain falling through a polluted atmosphere of concentration C_a is given by

$$C_d = Ca\ \Lambda^*/H_{wa} \quad (8.3a)$$

$$C_p = Ca\ \Delta \quad (8.3b)$$

in which the dimensionless gas and particle scavenging ratios, Λ^*as and Δ, respectively, vary from zero to unity, and H_{wa} is the air to water partition coefficient. The scavenging coefficient can be obtained from detailed modeling of precipitation scavenging of gases and aerosols (Pruppacher, Semonim, and Slinn, 1982). When precipitation scavenging and dry deposition of aerosols is considered, both the aerosol size distribution and pollutant distribution with respect to particle size are required.

The multimedia-compartmental model (eq. 8.1) can be expressed more conveniently in the following compact form:

$$\frac{d\mathbf{C}}{dt} = \mathbf{AC} + \mathbf{S} \quad (8.4)$$

in which **C** and **S** are the compartmental vectors of concentration and source strength per unit volume, respectively. **A** is the MCM parameter matrix whose parameters are defined as:

$$A_{ii} = \sum_{\substack{i-1 \\ i \neq j}}^{N} (-a_{ij}K_{ij}/V_i - Q_{ij}/V_i) + \xi_i k_i \tag{8.5a}$$

$$A_{ij} = a_{ij}K_{ij}H_{ji}/V_i + Q_{ji}/V_i, \qquad i \neq j \tag{8.5b}$$

The solution of the above model (Eq. 8.4) can be readily obtained either analytically or numerically depending on whether the parameter matrix **A** is a constant or a function of both concentration and time.

The MCM modeling approach can be used to determine pollutant partitioning in specific geographical locations. Such an approach was recently implemented in the MCM model of Cohen and Ryan (1984). An illustration of a test case simulation for the multimedia distribution of trichloroethylene (TCE) in the San Diego region (~ 400 km^2) is depicted in Figure 8–4. The presented test case (See Figure 8–4) is for a single trichloroethylene source of 430 gmole/hr.m^3 in the air phase, with the background concentrations based on field measurements (Su and Goldberg, 1976). The concentration-time profiles indicate a rapid increase in the concentration levels toward their steady state value. The effect of rain scavenging is most pronounced on the TCE level in the soil. Finally, after the elimination of TCE emissions, TCE is rapidly degraded in the atmospheric compartment, although it is more persistent in all other compartments. It is noted that dynamic concentration profiles as illustrated in Figure 8–4 can be directly employed in exposure calculations.

The above TCE example serves to illustrate the dynamic nature of pollutant partitioning in the environment. It is noted that the accuracy and therefore the applicability of the MCM approach are dependent upon accurate information on transport and partition parameters, the rate of various chemical transformations, and source emission data. In the absence of such knowledge, it is unlikely that the development of more complex models will lead to substantial improvements toward a priori predictions of pollutant fate and transport of pollutants in multimedia environments. Therefore, research efforts should first concentrate on providing information on intermedia transport processes and transformations.

SOURCES

General Considerations

The main driving force in any multimedia model is the total input of the chemical under consideration from sources to the system (Figure 8–3). Therefore, meaningful predictions from any multimedia model require the identification and quantification of all sources of the chemical. The pertinent information for each source should include:

Figure 8–4. Dynamic Partitioning of Trichloroethylene.

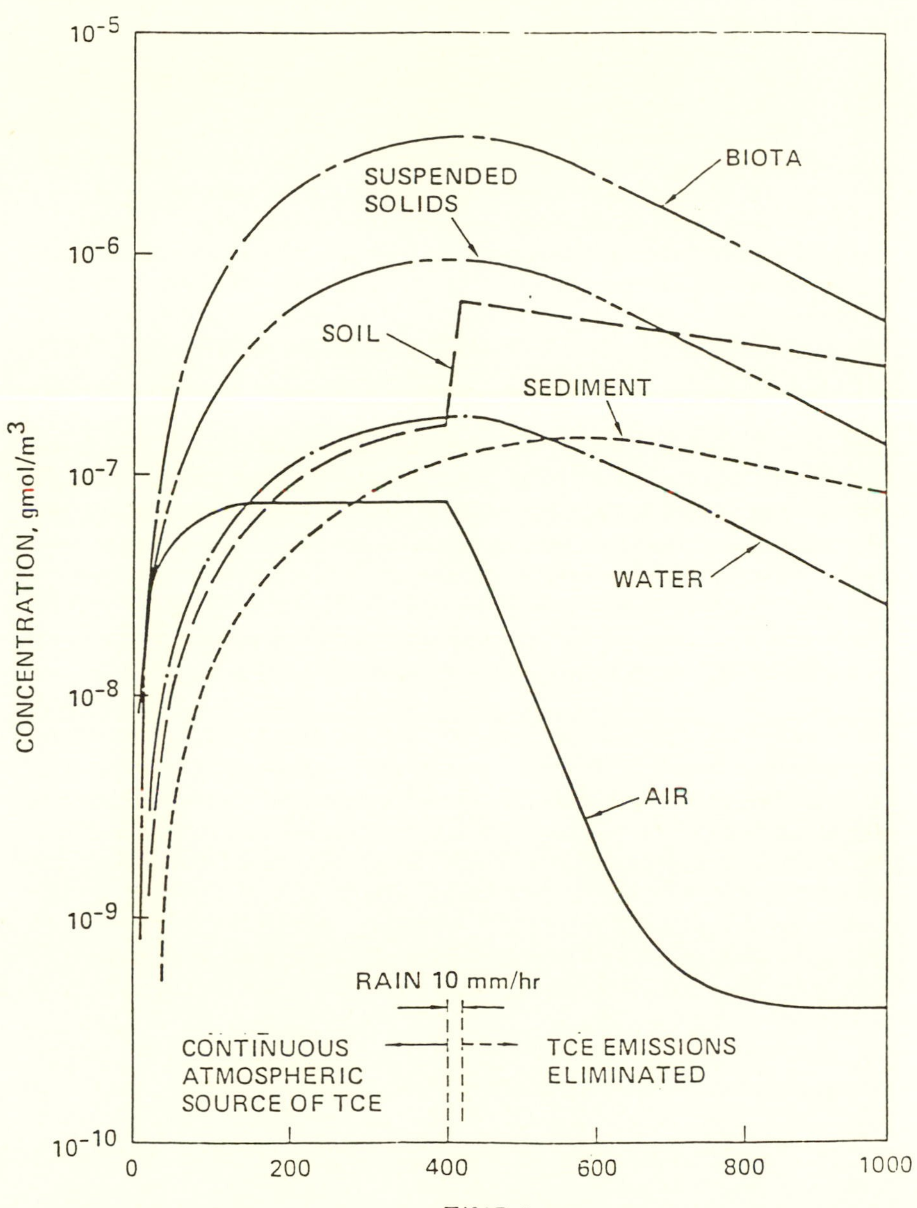

Table 8–2 Release of Chemicals to the Environment

A. To Atmosphere
1. Stack emission during manufacture
2. Fugitive volatilization losses, e.g., from leaks, storage tank vents, waste disposal, and treatment sites
3. Losses during use and disposal operations

B. To Water
1. Treatment of plant effluents at manufacturing and/or formulating plants
2. Spills during manufacturing (original and formulating) and distribution
3. Losses during transportation
4. Disposal after use

C. To Soil
1. Direct applications as an agricultural chemical or for vegetation or insect control
2. Land disposal, e.g., landfill or cultivation operations
3. Spills

1. total emission rate of the chemical to the particular environmental medium,
2. the physical state of the chemical (dissolved, particulate, or gaseous state),
3. the specific molecular form or speciation.

These data are needed for all stages of introduction of the chemical into the environment including losses during production, distribution, usage, and disposal.

The sources of pollution are many (see Table 8–2) and they vary in different locations with respect to the source strength and to physical and chemical characteristics of the released chemical. Although industrial wastes have received the most publicity, other sources such as agriculturally related contamination is equally important. Municipal landfills may also lead to the release of a complex collection of pollutants into the environment, and even releases from septic tanks can be significant.

Point source emissions cannot be accurately predicted and hence must be obtained from high quality emission inventories. Particulate emissions from industrial processes, for example, cannot be accurately predicted in most cases since they are largely determined by the efficiency of air pollution control devices. In many cases source-receptor modeling should aid in providing a better account of sources origin (Friedlander, 1973). The characterization of particle size is also an important factor since it has a significant effect on the transport of particulate pollutants.

Some dispersive emissions (e.g., trichloroethylene; Fuller, 1976) can be estimated from statistics on production and sales. More recently, O'Leary et al. (1983) have developed a methodology for estimating multimedia environmental loadings for chemicals even in the absence of manufacturing plant emission data.

The methodology draws on existing data bases for structurally similar compounds and knowledge of the manufacturing process.

Spills of hazardous substances are intermittent sources with the potential of leading to locally acute exposures. In most cases spills may account for only a small fraction of the total pollutant material balance. In certain cases, as in accidental spills of crude oil in aquatic environments, the dispersion of various crude oil components may occur over large areas (Fay, 1971). Consequently, this can lead to a devastating damage to the aquatic environment (Mann and Clark, 1978). Regardless of the size of chemical spills, they should be included through an established protocol and integrated into multimedia modeling efforts.

Pollutant release to the environment is also associated with disposal and treatment activities of hazardous waste. The disposal of hazardous compounds should not be confused with the interim or long-term storage of these compounds in manmade or natural reservoirs. The distinction between storage and disposal is a prerequisite to describing the total environmental cycle and multimedia transport of such substances.

Many of the hazardous wastes in the United States are deposited in lagoons, landfills, and open dumps. The accumulative effect of these disposal practices has resulted in the contamination of various local water supplies and air sheds (Dunlap, 1976; Shen and Swell, 1984). Consequently, unwarranted harmful health effects are continualy threatening various local communities. During chemical waste storage in the absence of biodegradation, losses to the environment proceed over time and ultimately may result in complete release. In reality the net effect of storage in leaky dumps is to provide a delay in release time, of the order of magnitude of the residence time in the dump. From a modeling viewpoint, the presence of dump sites requires the inclusion of sources from these sites, despite the possible slow release times.

Waste disposal by means of thermal degradation by pyrolysis or incineration, requires a careful evaluation of the trade-offs of one toxic substance for another (Sekulic et al., 1983). High temperatures during incineration may enhance the loss of volatile metals such as mercury and arsenic, and/or the formation of hazardous organic compounds such as polycyclic aromatic hydrocarbons. These emission streams then become part of the overall source allocation.

Research Needs

The various sources and modes of release of toxic chemicals into the environment (Table 8–2) must continue to be monitored in order to provide accurate source emission data that is needed for both pollutant budgets, source allocation models, and multimedia transport models. Concurrently, reliable methods should be developed for estimating sources of potential pollutants from current and planned chemical production and processing activities.

Figure 8–5. Pollutant Transport from the Atmosphere to Land and Water Compartments.

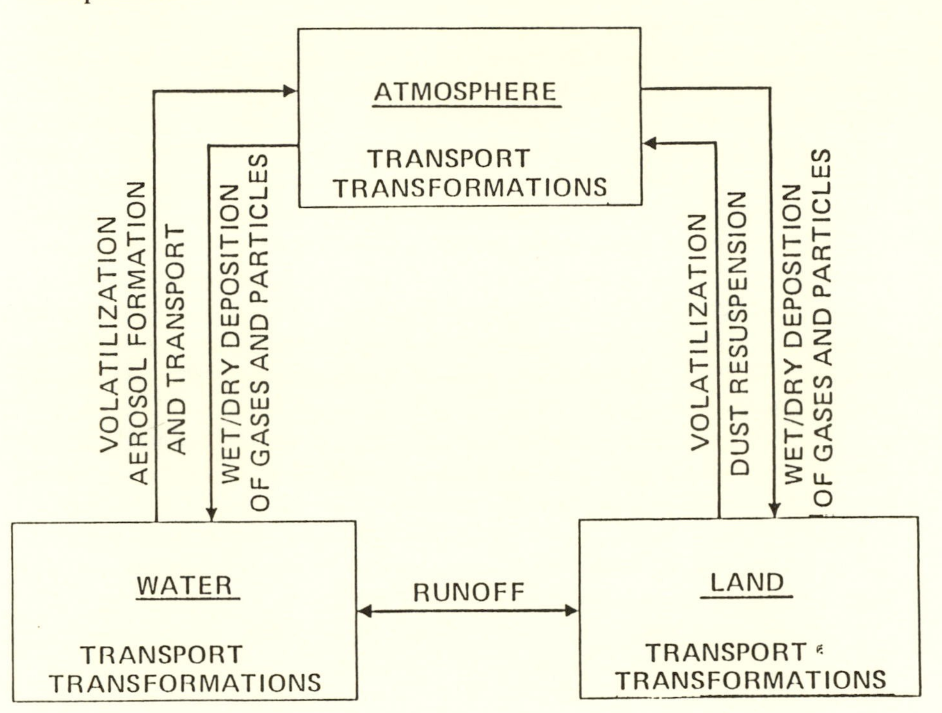

THE ATMOSPHERIC ENVIRONMENT

General Description

Many chemicals are introduced to the atmospheric environment by direct release, by evaporation from aquatic or terrestrial environments, from resuspension of dust and soil particles or through aerosol formation from aquatic bodies. After entering the atmosphere a pollutant is subjected to various meteorological phenomena such as wind and precipitation that transport it to various regions of the atmosphere and in some instances back to the aquatic and terrestrial environments (see Figure 8–5).

The atmosphere is a complex dynamic region of the environment. Air circulation, humidity, and temperature vary constantly throughout the various regions of the atmosphere. The atmosphere may be divided into several regions based on temperature (Heicklen, 1976). The lower region, called troposphere, is of primary interest in multimedia transport modeling, since most compounds enter the troposphere and are either degraded or transformed in that region. Few organics are likely to persist in the higher stratosphere because of the high energy ultraviolet radiation present there (McEwan and Phillips, 1975).

The movement of chemical pollutants into different regions of the atmosphere occurs at different rates (Levy, 1974; Neely, 1977a; Singh, 1979; Derwent, 1982). The convective half-life associated with movement from the troposphere to the stratosphere is of the order of several decades (about 10–30 years). The convective half-life for transport between the northern and southern troposphere is of the order of one year. The half-life associated with mixing within the troposphere is about one month. Mixing time scales on a smaller geographical scale are much shorter, of the order of a few hours to several days.

There are numerous atmospheric flow models that can be used to treat convection and dispersion processes throughout the atmospheric region (Bonazountas and Fiksel, 1982; Seinfeld, 1975; Turner, 1970; Start and Wendell, 1974; Hanna, 1973; Culkowski and Patterson, 1976; Nieuwstadt and Van Dop, 1982; and references cited by these authors). A review of the dynamics of the atmosphere is beyond the scope of this study. This area of research has reached a degree of sophistication that is probably beyond the current needs of multimedia modeling efforts. In contrast, information on transformation reactions of organics in the atmosphere, and reliable prediction methods of intermedia transport processes, such as dry-deposition and precipitation scavenging in the field are still underdeveloped (Friedlander and Pruppacher, 1981).

The lifetime of pollutants in the atmosphere is strongly affected by their chemical reactivity in both the gas and aerosol phases. Compounds in the atmosphere may react by at least three types of processes (McEwan and Phillips, 1975; Lyman et al., 1982; Darnall et al., 1978; Finlayson and Pitts, 1976; Hampson and Garvin, 1977): 1) direct photoloysis by sunlight, 2) electronic energy transfer, and 3) reaction with various species normally present in the atmosphere. Very long reaction lifetime will lead to stratospheric sinks with dry and wet deposition becoming important (Bjorseth and Lunde, 1979). At short residence times, the nature of any resulting reaction products may become important in determining the resulting concentrations in urban and industrial areas.

The fate of organic chemicals in the atmosphere is particularly complex, since many organics are present in both gaseous and particulate form in the atmosphere (Cantreels and Canwenberghe, 1978). Furthermore, photochemical and thermal reaction can transform the parent compound into more or less toxic daughter compounds (Pitts et al., 1978; Cliath and Spencer 1972). Additionally, significant reintroduction of organics from soil and water bodies into the atmosphere by volatilization of even low vapor pressure compounds may occur (Spencer and Cliath 1975; Glotfelty and Caro, 1975).

Due to the above difficulties as well as an inadequate source data base, atmospheric mass balances are available for only several organic compounds such as DDT (Woodwell et al., 1971; Cramer, 1973; Harrison et al., 1970).

PCB (McClure, 1976), B(a)P (Abrott et al., 1978), Trichloroethylene (Cohen and Ryan, 1984), chloroform (Khalil et al., 1983; Abrott et al., 1978), and formaldehyde (Thompson, 1980).

Research Needs

The formulation of successful multimedia transport models will require further basic research work in the following intermedia (or interfacial) processes: 1) aerosol and organic vapor wet and dry deposition processes, and 2) size and chemical composition distribution for various organic chemicals in the aerosol phase. Further research also needs to be conducted on transformation reactions of organics and the sorption of organic vapors by aerosols, as well as on gas-surface transformation reactions.

Multimedia models should include both the aerosol and gas phase. Emphasis should be placed on organic compound for which substantial information is already available (e.g., B(a)P). The mass balance should take into account chemical reactions, environmental flows, and various intermedia transport processes. Prototype mass balances should also be conducted, for compounds that are primarily in the vapor phase (e.g., benzene and trichloroethylene), primarily in the aerosol phase (e.g., B(a)P and other polynuclear aromatic hydrocarbons), and compounds existing in both phases. These evaluations will help to identify the deficiency of various multimedia models and the precise areas where more basic research information, is needed.

After the prototype models are developed, the next step is to conduct multimedia field studies to validate the models. This testing, will require measurement of organic compounds in both the gas and aerosol phases. Inventory data on atmospheric organics should be established. Additionally, there is a need to develop and validate methods to integrate deposition velocities across spatial inhomogeneities in terrain, and in vegetative covers. Concurrently, models for simultaneously investigating meteorological transport, diffusion, and dry-deposition removal will have to be improved and validated for incorporation in an overall multimedia scheme.

THE TERRESTRIAL ENVIRONMENT

The Soil Compartment

The soil is a complex multiphase system that consists of soil solids, soil-air, soil-solution, and biota. The chemical environment of the soil (pH, redox potential, ionic composition of the soil solution) is a sensitive function of its recent history, temperature cycles, and humidity cycles (Alrichs, 1972; Kohnke, 1968).

The fate of various chemicals in the soil is governed by the following factors:

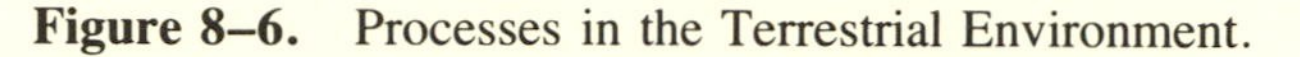
Figure 8–6. Processes in the Terrestrial Environment.

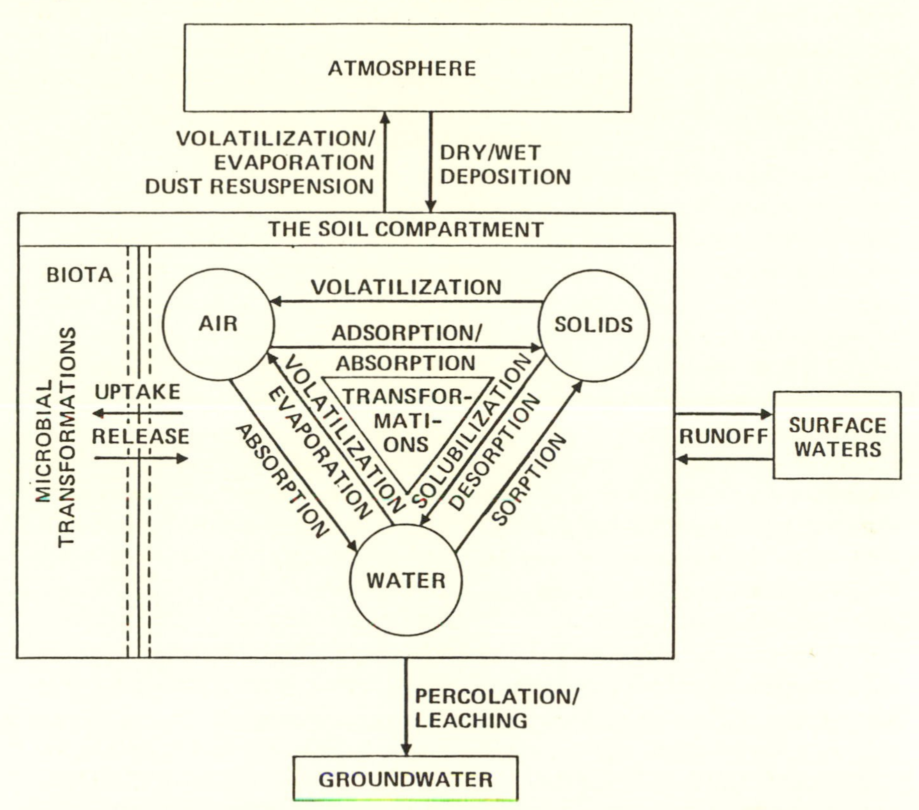

1) sorption/desorption and the associated thermodynamics, 2) physicochemical and biological transformations, 3) photodegradation and volatilization, 4) transport phenomena associated with leaching and exchange processes, and 5) properties and characteristics of soils. A schematic representation of the above processes in the soil is shown in Figure 8–6.

Many contaminants are known to associate strongly with soil and sediment via adsorption (Hamaker and Thompson, 1972; Karickhoff, 1980, 1981; Karickhoff et al., 1979). Most soils contain some organic matter to which hydrophobic molecules tend to adsorb. Also, some soils have an ion exchange binding capacity that is effective in holding cationic forms of toxic heavy elements. Much of the available work on sorption of organic chemicals has been related to pesticides, and excellent reviews of the subject are available elsewhere (Bailey and White, 1964; Farmer 1976; Hamaker and Thompson, 1972; Kenega and Goring, 1980). Sorption studies to date have yielded important information on adsorption kinetics and adsorption isotherm correlations. For example, the soil organic content, and the water/octanol partition coefficient have been identified as important correlating variables (Karickoff, 1981; Kenaga and Goring, 1980). Unfortunately the available correlations are of insufficient generality and may lead to significant errors (Cohen and Ryan, 1984) when extrapolated to different soils, or compounds on which the correlations were not directly based.

Chemical transformations of pollutants in the surface and subsurface soil include photodecomposition, chemical hydrolysis, chemical oxidation, and biotic reactions. For most pollutants, particularly organic compounds, biotic reactions are more important than abiotic reactions. In the aerobic zone (surface soil zone) bacteria may oxidize many of the organic constituents. Under the usual anaerobic conditions in the deep soil region metabolism of most hydrocarbons will not occur. In contrast, chlorinated organic compounds seem to be more likely to be attacked in an anaerobic environments than under aerobic ones.

Pollutant movement in the soil compartment can occur by diffusion and leaching (Hamaker, 1975; Farmer et al., 1976; Farmer, 1976; Bailey and White, 1964). Leaching is dependent on water movement in the soil and the adsorption/desorption isotherms for the contaminant. Correlations for sorption of unionized organic compounds and organic matter content of soils are reasonably well established. Also, the modeling of leaching in laboratory columns has been fairly successful. The application of leaching models in the field however, have met with only limited success due to spatial variability of soil characteristics, lateral diffusion, and water migration.

The exchange of pollutants between the soil and the atmosphere is determined largely by volatilization (Hamaker, 1972; Farmer et al., 1976; Spencer and Cliath, 1975). Volatilization from soils is the predominant loss mechanism for many low solubility organic contaminants. Additionally such volatilization processes are governed by diffusive transport in the soil and by aerodynamic factors operating at the air/soil interface (Farmer et al., 1972). The transport of pollutants to the atmosphere may also occur by the resuspension of particulate matter (Pruppacher et al, 1982; Travis, 1975). Pollutants may also be transported to water bodies through runoff or through leaching to groundwater.

Research Needs

Current prediction methods for pollutant adsorption on soil are often unsatisfactory. Thus there is certainly a need to gather more isotherm data and to continue the correlation of adsorption isotherms for groups of similar compounds. Additionally, there is a need for a reliable data base, and of prediction models of biodegradation reactions of various chemicals in the soil.

It is known that sorption of certain low molecular hydrocarbons (e.g., trichloroethylene) may continue for weeks and months on clay soils (Vilker, 1984). Laboratory sorption experiments, however, are generally carried out at fixed temperature and moisture conditions, and over short time scales. The above suggests that data on long-term adsorption including the effect of moisture and temperature are needed. This information is of particular importance for long-range forecasting of pollutant transport in the soil compartment.

There is a need to further develop and test models for predicting the volatilization of various organic chemical from soils which include simultaneous movement of chemicals toward the surface by diffusion and convective transport due to evaporating water, while at the same time considering the effect of chemical degradation of the compound in the soil. The effect of temperature gradients in the soil including diurnal reversal of the temperature gradient in the upper soil layers, and soil moisture cycles should also be included in the model formulation.

Current available models of pollutant transport in the unsaturated soil matrix (Bonazountas, 1983) account for soil-water/soil-air or soil-solids/soil-water intermedia processes but often neglect soil-solids/soil-air processes. Recent multimedia modeling efforts of soil processes have emphasized the compartmental approach. Although compartmental soil transport models such as the SEASOIL model (Bonazountas and Wagner, 1981) are extremely useful, they require extensive field data for model calibration. Consequently, compartmental soil models may lack a priori predictive capability especially for heterogeneous soils.

It is important to realize that the soil matrix cannot be modeled effectively as an array of well-mixed compartments. This stems from the inherent nonuniform characteristics of the soil in which transport processes take place through primarily diffusive processes or convection at low Reynolds number flows. This results in time dependent and highly nonuniform pollutant concentration profiles. Clearly, multimedia soil transport models must be time dependent and at least one-dimensional. Furthermore, multimedia soil transport models should simultaneously treat transport processes occurring in heterogeneous multimedia soil environments (i.e., soil-solids, soil-water, soil-air, and biota compartments) and the interactions of the soil with the atmospheric phase and groundwater.

Models of soil transport that include runoff are available at various levels of complexity (Carsel, 1980; Crawford and Donigian, 1973; Knisel, 1980; Johanson et al., 1980; Johanson, 1983). Such models can be employed in the formulation of multimedia transport models.

The Plant/Soil and Plant/Atmosphere Interfaces

General Description

Plants are the basic food components in the food chains to humans hence their role as mobilizers and sinks of pollutant is particularly important. Plants may absorb and transport pollutants from the soil or absorb pollutants from the atmosphere (Nash, 1974; Wallace and Berry, 1981).

The zone beginning at the root surface and extending far into the bulk of

Figure 8–7. Exchange Processes at the Plant/Soil Interfaces.

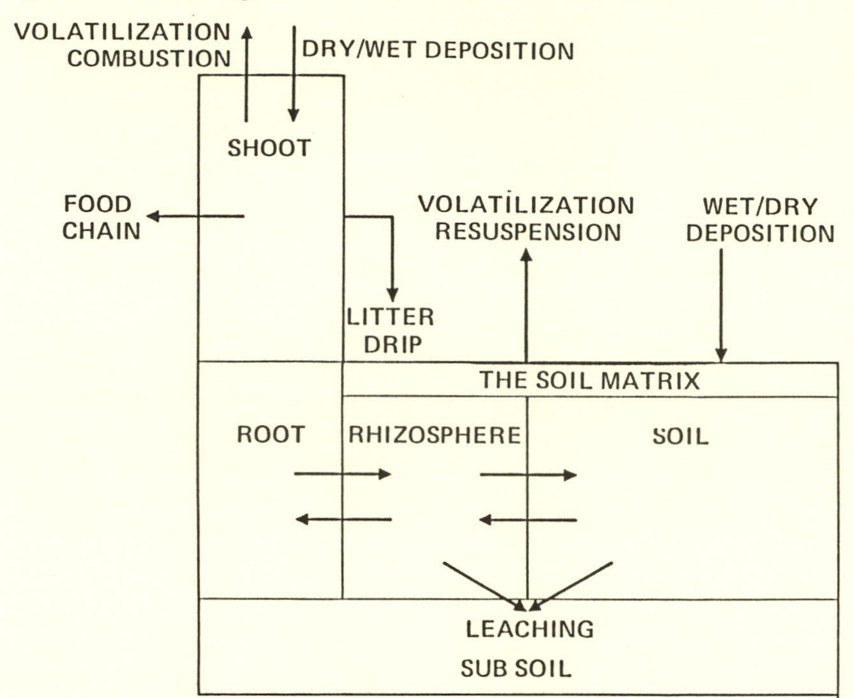

the soil, with its biota (i.e., the rhizosphere) is very active microbially. In this zone, transformations and mobilization of mineral nutrients and pollutants occur. Pollutants may be degraded before becoming available to plants. Pollutants may also deposit on shoots of plants, and they may be taken up through plant leaves. A schematic of the interactions of the plant compartment with the atmosphere and soil environments is shown in Figure 8–7.

Research Needs

Models for predicting metabolism and/or transport for various chemicals through the plant/soil interface need to be developed. Such models will be essential in order to avoid the impossible task of empirically evaluating all possible pollutants. The basic questions that have to be answered are:

1. What are the pathways by which pollutants move between the soil-plant, plant-air interfaces?
2. What are the driving forces and flux direction?

Specific topics should include:

1. Influence of root, rhizosphere, and soil on transformations and uptake of pollutants;
2. Characterization of the partitioning and bioaccumulation of pollutants within the plant in relation to food chain incorporation;
3. Volatilization of pollutants from plants and rhizosphere to the atmosphere;
4. Foliar uptake and transport of pollutants.

Groundwater

Pollutants can find their way into the nation's groundwater via injection wells, land fills, surface mining, and rapid infiltration (Pye and Quarles, 1983). Landfills are often located in the vicinity of populated regions and are possible sources of pollution leaking into groundwater for many years. Surface mining is also a potential source of shallow aquifier contamination, because mining operations are often below the shallow water table.

Organic pollutants are persistent in underground water. Since groundwaters are normally anaerobic, oxidation is therefore unlikely. Also, if bacteria are present they must be anaerobs or facultative aerobs, and both generally degrade organic matter more slowly and less completely than aerobs (Metcalf and Eddy, 1979).

There are numerous models that deal with the contamination of groundwaters (Pinder, 1984; Bonazountas 1983; Mackay et al., 1985). These models however, are highly dependent on the availability of accurate adsorption/desorption isotherms and the associated diffusional resistance, as well as chemical and biological transformations. The inclusion of pollutant movement in groundwater in an overall environmental multimedia model is difficult since groundwater pollution problems are generally site specific, requiring accurate description of aquifer physical characteristics. Consequently, a high degree of spatial resolution is required, and multimedia models that include groundwater movement will ultimately also become site specific.

THE AQUATIC ENVIRONMENT

General Description

The introduction of chemicals into the aquatic environment admits the possibility of loss of aquatic life or productivity, the creation of toxic hazards for human drinking and contacting the water, and the creation of unpleasant odors or appearance that can reduce the desirability of aquatic resources.

Chemicals that are introduced into the aquatic environment are dispersed and transformed by various physical, chemical, and biological processes, as shown schematically in Figure 8–8. The chemicals may partition between the

Figure 8–8. Pollutant Transport and Transformation in the Aquatic Environment

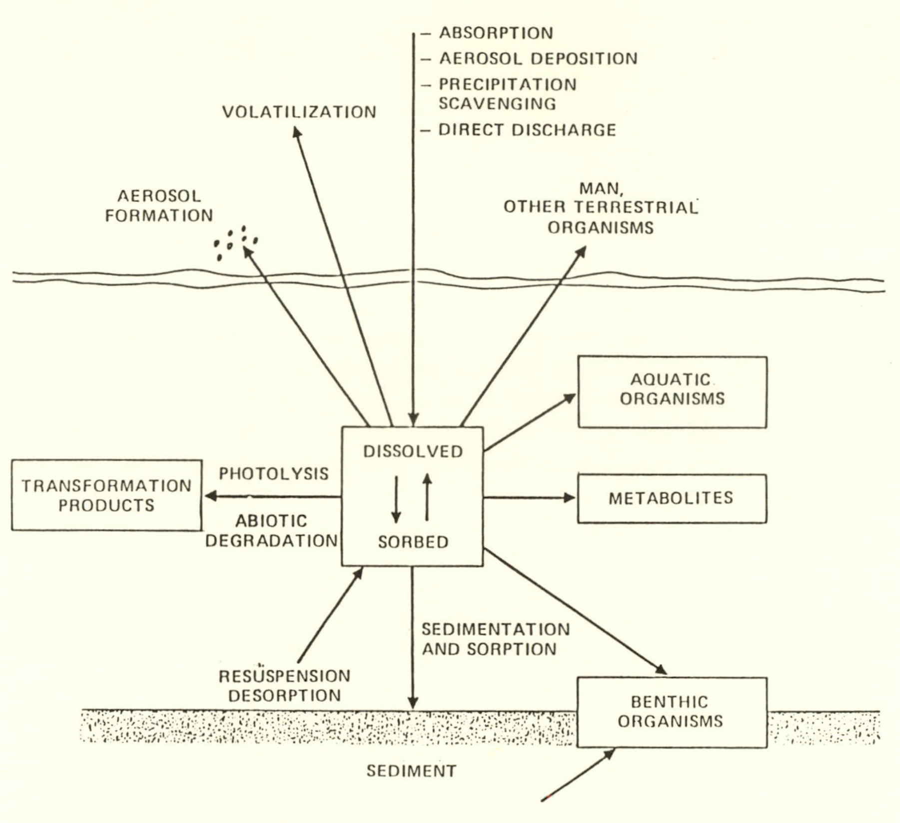

aqueous phase, suspended solids, and sediment by adsorption, and be convected within the aqueous phase and across the water/atmosphere interface. Chemicals may leave the water phase and enter the atmosphere by volatilization (Cohen et al., 1978; Cohen, 1983) or by aerosol formation (Blanchard, 1982). Transport of material from land to water may also take place via runoff, groundwater movement, snow melt, wind erosion, and deposition (Novotny and Chesters, 1981). Pollutants may enter the aqueous phase from the atmosphere by precipitation scavenging, dry deposition of aerosols, and gas absorption (Pruppacher et al., 1982; ref. 99). The transport of chemicals in the aquatic environment is also mediated by biota, which takes up chemicals from the water and may accumulate or metabolize them. Chemicals may also be subjected to various transformation processes such as photolysis, oxidation, reduction, hydrolysis, and microbial degradation.

The rates of photolysis in natural water may differ significantly from those measured in pure water because of the presence of naturally occurring light absorbers, quenchers, or sensitizers (Miller and Zepp, 1979). Naturally occurring materials such as humic and fulvic acids which have high optical densities, may absorb sunlight and effectively screen chemicals from being photolyzed. The presence of particulate matter may also hinder photolysis due to light scattering by particles.

Two other major transformations are oxidation and hydrolysis (Smith et al., 1978). Hydrolysis is an important degradation reaction for many organic pollutants (Mabey and Mill, 1978). Rates of hydrolysis are generally second order reactions with the rate constant highly dependent on pH. Oxidation reactions occur through the interaction of dissolved pollutants with free radicals.

Heterotropic microorganisms, found in virtually all aquatic environments (water phase and sediment), are capable of degrading many xenobiotic compounds as well as naturally occurring substances and thus constitute an important mechanism for removal or transformation of contaminants. Environmental conditions influence the number of microorganisms present, rate of microbial degradation, and metabolic pathways (Bourquin and Pritchard, 1979). The rates of microbial transformations also depend on temperature, dissolved oxygen concentration, and upon the magnitude (high versus low concentration) and continuity (acute versus chronic) of the contaminant source (Wright and Hobbie, 1965; Ward and Brock, 1976; Herbes and Schwall, 1978).

The dispersion of chemicals in the aquatic environment is dominated by processes of convection and turbulent diffusion. Hydrodynamic dispersion processes are complex, but predictable from either numerous theoretical models, or in specific sites through the use of brine, dyes, and other conservative tracers. The degree of difficulty in predicting dispersion processes increases in the following order:

1. Steady state unidirectional flow (river during constant flow);
2. Nonsteady state unidirectional flow (river after a storm);
3. Steady state multidirectional flow (lake, sea);
4. Unsteady state multidirectional flow (tidal estuary).

Dispersion due to wave action and mesoscale circulation dominates the open ocean environment (Robinson, 1976; ref. 99). At high energy coastlines with a rocky or sandy shore, strong surf action and frequent storms will normally disperse waterborne pollution rapidly. In low energy coastline sedimentation is more prevalent and buried sediment will occasionally uncover and disperse in storms. In estuarine waters, fresh water flowing into salt water compounded by tidal mixing can result in long retention times in the absence of stormy weather or strong offshore current. At low degree of fluid mixing, as in small lakes, sedimentation of toxic materials is an important process that leads to the enrichment of sediment with toxic chemicals.

Uptake and release of contaminants from suspended particles is affected by the hydrodynamics of the water phase, and biological activity (Lee, 1979; Lehmann, 1977). Mixing processes within the sediment and between the sediment and the overlying water, in both large contained water bodies and flowing streams, often govern the environmental significance of sediment-associated contaminants (Lee, 1979; Lehmann, 1977; Lerman and Ligtzke, 1975). Most of the existing aquatic transport models provide a parametric treatment of water-sediment interactions (Onishi and Wise, 1979). The model parameters in the majority of the available models have been adjusted to give a close fit between predicted and observed concentrations. Consequently, the utility of such models as a priori forecasting tools is often limited.

Historically most applications of contaminant transport in the aquatic environment have been site specific. There are models to describe the spreading of nonpoint sources (Onishi and Wise, 1979; Donigian and Crawford, 1976; Smith et al., 1978). There are also unified models that include groundwater and runoff processes such as the UTM family of models (Burns et al., 1981) and the HSPF modeling system (Johanson, 1983). More recently, compartmental models have been developed to provide prediction of contaminant behavior in the aquatic environment under widely varying environmental conditions (Burns et al, 1981; Park et al., 1982).

Research Needs

There exist a number of models of ocean, lake, and river hydrodynamics that can be used to model the dispersion of contaminants in the marine environment. Such models can be coupled with other single-medium transport models, or they can be used as guides in compartmentalizing the aqueous compartment, and subsequently combined with an overall multimedia description of pollutant behavior in the environment. There remain, however, several key areas of intermedia transport and transformation processes that are not well understood.

Solubility and air/water partition coefficient data for pollutants in natural waters are often lacking, although the experimental methodology for determining these data is available (Mackay et al., 1979). Adsorption isotherms for suspended solids and sediment can be predicted from available correlations, but the reliability problem with these correlations still remains. The research needs in this area are similar to the ones already outlined for the soil compartment.

Prediction methods of environmental volatilization and absorption rates of pollutants are now available (Cohen, 1983; Mackay and Yeun, 1983; Brutsaert and Jirka, 1984). The available predictions do not consider the effect of naturally occurring soluble surface active chemicals (Parker and Barsom, 1970). Also lacking is information on the effect of microbial activity, in the sea surface layer, on gaseous exchange and photolysis. Additionally, the transport of pollutants from aquatic bodies to the atmosphere through the formation of aerosols (MacIntyre, 1974; Blanchard, 1982) is inadequately documented at this time.

More information is needed on the processes of sedimentation and resuspension which are particularly important for persistent organics. Predictive models for sediment removal of both soluble and particle-bound pollutants in oceans, lakes, and rivers should then be validated using knowledge of the above intermedia processes.

There is a need for a comprehensive data base on the transformation of organic chemicals in natural waters. The data base should include: a) hydrolysis by water, H_3O^+pl, OH $^-$ms ; b) photolysis by direct adsorption of sunlight; c) oxidation by free radicals and singlet oxygen; d) microbially mediated transformations; and d) oxidation and reduction effect of dissolved oxygen content and the presence of various catalysts including metals, acids, bases, and ultraviolet light as well as temperature effects.

Finally, the inclusion of the biota compartment will require the formulation of models for uptake by biota, given bioconcentration factors from either laboratory measurements or from correlations with the octanol/water partition coefficient, as well as kinetic uptake parameters (Veith et al., 1979; Neely et al., 1974; Chiou et al., 1977; Neely, 1979). The approach employed in the PEST model (Park et al., 1982) for example may serve as a guide for modeling biota compartments in overall multimedia scheme.

MICROCOSM MODELS

Microcosm studies have been designed with the intention of reproducing on a small scale various combined processes of biological systems in a model environment (Pritchard, 1982). Microcosms, which may be either terrestrial, aquatic, or both, are intended to serve as surrogates of the real world by incorporating the main aspects of toxic chemical pathways through the ecosystem (Cole and Metcalf, 1974; Cole et al., 1976; Gillet and Gile, 1976; Litchtenstein et al., 1978; Metcalf, 1977b).

Microcosm studies provide indication of the environmental partitioning as well as estimates of the persistence of various chemicals (Gard, 1982). Microcosms may serve as useful tools in screening various processes under controlled laboratory conditions of temperature, humidity, light, water balance, and the introduction of various biotic and nonbiotic compounds (Draggan, 1976a, 1976b; Holm et al., 1982). Microcosms are probably most useful in simulating the exposure of various biological species in a multi-species communities subjected to contact with one or more toxic pollutants, through multimedia exposures (Sugiura et al., 1976; Metcalf, 1977a). In this manner routes of exposure can be identified.

The major flaw of microcosms is that they are not self-sustaining and have impermeable boundaries. Therefore, the dynamic behavior of the environment cannot be duplicated and hence the extrapolation of average or even maximum concentrations or exposure levels from microcosm data to the real environment must be made with great reservations (Pritchard, 1982). The scaling procedure of microcosm data to the environment is not well established and one must often rely on heruistic set of scaling rules (Hill and Weigert, 1980). There is a need for research to establish the degree of permitted extrapolation of small microcosms data to more complex microcosms and to actual environmental systems. Such research will aid in establishing the utility of microcosms in validating multimedia models of pollutant transport and accumulation in the environment.

HAZARDOUS WASTE TREATMENT AND DISPOSAL SITES: A MULTIMEDIA PROBLEM

General Description

Chemical waste disposal and treatment sites represent a unique multimedia pollution problem. Hazardous waste from both disposal and treatment sites are rarely totally isolated from the surrounding environment. Therefore, toxic chemicals do reenter the atmosphere, soil, and groundwater, albeit at a slow rate compared with direct emissions from industrial or domestic waste streams. Subsequently, these chemicals enter a multimedia transport cycle since they can be convected through the air phase and deposited either through dry or wet deposition onto soil, vegetation, or water bodies (Shen and Swell, 1984; Dunlap, 1976; Thibodeaux et al., 1984). Chemicals that are leached into underground water aquifers may contaminate drinking water sources directly, or they may contaminate larger natural water bodies, from which they can be transported back into the atmospheric phase through volatilization, or aerosol formation, or they may be adsorbed by the underlying sediment. Clearly, in order to assess the potential environmental impact of hazardous waste disposal or treatment practices a multimedia description of such hazardous waste sites is needed.

Existing state and federal laws have severely limited the amount of wastes that can be disposed of in surface waters and the atmosphere. Burial in the ground has therefore become a commonly used option for the disposal of hundreds of millions of tons of wastes produced each year. Present regulations and standards of hazardous waste sites emphasize protection against surface and underground water supplies. Air pollution problems associated with hazardous waste facilities have not been adequately addressed (Shen and Swell, 1984). Although current technology can be used to retard fugitive emissions from waste sites, the emissions problem still remains.

Research Needs

A multimedia-compartmental approach to the modeling of pollutant transport at hazardous waste sites should provide a first-level assessment of environmental impact associated with a particular site. Multimedia models making use of detailed transport equations accounting for both temporal and spatial variations are, however, needed in order to provide detailed information for the monitoring and design of hazardous waste sites.

SUMMARY

The formulation of environmental models designed to assess the presence, distribution, and effect of chemicals in the environment is an extremely complex, multi-parameter problem. One of the problems in designing effective predictive models is in minimizing laboratory and field data required for model validation. Therefore the design of an accurate multimedia model necessitates a thorough knowledge of each of its individual environmental compartments and their mutual interactions.

An integrated multimedia systems approach to forecasting pollution hazards is needed. A multimedia approach would identify the critical informational elements necessary for a comprehensive environmental pollution abatement program. A multimedia approach would also reveal potential environmental "hot spots" of toxic chemicals.

It is accepted that it will be impossible to test all potential pollutants in detail. Therefore, there is a need for a program that couples multimedia models and multimedia monitoring studies with "benchmark" chemicals which are widespread in the environment (ref. 117, 118). The physical and chemical properties of these chemicals should be examined in detail as well as their transformation reactions in the various phases under different environmental conditions. Such integrated studies will enhance efforts to identify the complex pollutant pathways leading from pollution sources to uptake by humans.

At the present time, multimedia modeling programs are in an embryonic stage. Although the existing models represent a step in the right direction, these models are oversimplified with an only approximate treatment of intermedia transport processes. The coordination and amplification of the above research activities is most essential for progress toward assessing the environmental impact of emerging new technologies, and hazardous waste treatment and disposal practices.

At the current state of the art of multimedia research, multimedia-compartmental (MCM) models should provide a first stage for the design of monitoring activities and risk assessment. MCM models should provide a focal point for sorting out pertinent information regarding pollutant behavior in the environment. In this manner MCM models will become a valuable tool for guiding future research efforts.

While there is a strong need to maintain research efforts on processes within various environmental compartments, research programs on the multimedia nature of environmental pollution should be established. This goal cannot be achieved through a simple reorganization of administrative structure of governmental agencies since it must include a research effort in multimedia pollution as a discipline. This goal will require the establishment of recognized multidisciplinary research groups, centers, or institutes of multimedia pollution research.

REFERENCES

1. Abrott, T. J., M. J. Barcelona, W. H. White, S. K. Friedlander, and J. J. Morgan, "Human Dosage/Emission Source Relationships for Benzo(a)pyrene and Chloroform in the Los Angeles Basin," Special Report to US/EPA, Keck Laboratories, Caltech (1978).
2. Addison, R. F., S. Paterson, and D. Mackay, "The Predicted Environmental Distribution of some PCB Replacements," *Chemosphere* 12: no. 6, 827–34 (1983).
3. Alrichs, J. L., "The Soil Environment," in *Organic Chemicals in the Soil Environment,* C.A.I. Goring and J. W. Hamaker (eds.), vol. 1, Marcel Dekker, New York (1972).
4. Bailey, G. W. and J. W. White, "Review of Adsorption and Desorption of Organic Pesticides by Soil Colloids, with Implications Concerning Pesticide Bioactivity," *Agr. Food Chem.* 12:324 (1964).
5. Behar, J. V., E. A. Schuck, R. E. Stanley, and G. B. Morgan, "Integrated Exposure Assessment Monitoring," *Environmental Science and Technology* 13:34–39 (1979).
6. Bjorseth, A. and G. Lunde, "Long-range Transport of Polycyclic Aromatic Hydrocarbons," *Atmos. Environ.* 13:45 (1979).
7. Blanchard, D. C., "Aerobiology and Water-Air Exchange," ASRC Report, 1979–81, Atmospheric Sciences Research Center, State University of New York at Albany (1982).
8. Bolin B., A. Bjorkstrom, and K. Holmen, "The Simultaneous use of Tracers for Ocean Circulation Studies," *Tellus* 35B:206–36 (1983).
9. Bonazountas, M. and J. Wagner, "SEASOIL: A Seasonal Soil Compartment Model," Office of Toxic Substances, United States Environmental Protection Agency, Washington, D.C. (1981).
10. Bonazountas, M. and J. Fiksel, "ENVIRO: Environmental Mathematical Pollutant Fate Modeling Handbook/Catalogue," EPA No. 68–01–5146, Arthur D. Little, Cambridge, MA (1982).
11. Bonazountas, M., "Soil and Groundwater Fate Modeling," in *Fate of Chemicals in the Environment,* R. L. Swann, and A. Eschenroeder (eds.), ACS Symposium Series No. 225, American Chemical Society (1983).
12. Bourquin, A. W. and P. H. Pritchard, "Microbial Degradation of Pollutants in Marine Environments Workshop," U.S. EPA, Ecological Research Series, EPA, 600/9–79–012, Cincinnati, OH (1979).
13. Browman, M. G., M. R. Patterson, and T. J. Sworski, "Formulation of the Physicochemical Processes in the ORML Unified Transport Model for Toxicants (UTM-TOX) Interim Report," ORML/TM–8013, Oak Ridge Natl. Laboratory, Oak Ridge, Tennessee (1982).
14. Brutsaert, W. and G. H. Jirka, (eds.), *Gas Transfer at Water Surfaces,* D. Reidel, Boston (1984).

15. Burns, A. L., P. M. Cline, and R. R. Lassiter, "Exposure Analysis Modeling System (EXAMS): User Manual and System Documentation," Environmental Research Laboratory, U.S., Environmental Protection Agency, Athens, Georgia (1981).
16. Burton, G. B., "The Exposure Commitment Method in Environmental Pollutant Assessment," *Environmental Monitoring and Assessment* 1:21–36 (1981).
17. Cantreels, W. and R. Van Canwenberghe, "Experiments on the Distribution of Organic Pollutants Between Airborne Particulate Matter and the Corresponding Gas Phase," *Atmos. Environ.* 12:1, 133–141 (1978).
18. Carsel, R. F., *Pesticide Runoff Simulator User's Manual,* Computer Sciences Corporation (1980).
19. Chawgla R. C. and M. M. Varma, "Pollutants Transfer Between Air, Water and Soil Criteria for Comprehensive Pollution Control Strategy," *J. Environmental Systems* 11(4):363–74 (1982).
20. *Chemical Engineering News,* editorial, "Panel Suggests Management Changes at E.P.A.," June 4, 33 (1984).
21. Chiou, C. T., V. H. Freed, S. W. Schmedding, and R. L. Kohnert, "Partition Coefficients and Bioaccumulation of Selected Organic Chemicals," *Environ. Sci. Technol.* 11:475–78 (1977).
22. Cliath, M. M. and M. Spencer, "Dissipation of Pesticides by Volatilization of Degradation Products, I. Lindane and DDT," *Env. Sci. Technol.* 6:910 (1972).
23. Cohen, Yoram, W. Cocchio, and D. Mackay, "Laboratory Study of Liquid Phase Controlled Volatilization Rates in the Presence of Wind-Waves," *Env. Sci. Tech.* 12:553 (1978).
24. Cohen, Yoram, D. Mackay, and W. Y. Shiu, "Mass Transfer Rates Between Oil Slicks and Water," *Can. J. Chem. Eng.* 58:569–75 (1980).
25. Cohen, Yoram, "Compartmental Modeling of Environmental Transport," Report 81–1, National Center for Intermedia Transport Research, University of California, Los Angeles, November (1981).
26. Cohen, Yoram, "Mass Transfer Across a Sheared, Wavy Air-Water Interface," *Int. J. Heat and Mass Transfer,* 26:1,289–297 (1983).
27. Cohen, Yoram and P. A. Ryan, "Multimedia Modeling of Environmental Transport: Trichloroethylene Test Case," *Environ. Sci Technol.* 9:412 (1984).
28. Cohen, Yoram and P. A. Ryan, "Environmental Partition Coefficients," NCITR Report No. 84–2, National Center for Intermedia Transport Research, University of California, Los Angeles (1984).
29. Cole, L. K. and R. L. Metcalf, "Terrestrial Aquatic Laboratory Model Ecosystem for Pesticide Studies" in *Proceedings of the Symposium on Terrestial Microcosms and Environmental Chemistry,* J. W. Gillet and J. M. Witt (eds.), Corvallis, Oregon, June (1974).
30. Cole, L. K., R. L. Metcalf, and J. R. Sanborn, "Environmental Fate of Insecticides in Terrestrial Model Ecosystems," *Intern. J. Environ. Stud.* 10:7 (1976).
31. Conway R. A., *Environmental Risk Analysis for Chemicals,* Van Nostrand Reinhold, New York (1982).
32. Conway, R. A., "Introduction to Environmental Risk Analysis," in *Environmental Risk Analysis for Chemicals,* R. A. Conway (ed.), Van Nostrand Reinhold, New York (1982).
33. Cramer, J., "Model of the Circulation of DDT on Earth," *Atmos. Environ.* 7:(1973).
34. Crawford, M. and R. S. Donigian, Jr., "Pesticide Transport and Runoff Model for

Agricultural Lands,'' E.P.A. 6602–74–013, Office of R and D, U.S. Environmental Protection Agency, Washington, D.C. (1973).

35. Culkowski, W. M. and M. R. Patterson, ''A Comprehensive Atmospheric Transport and Diffusion Model,'' ORNL/NSF/EATC–17, Oak Ridge National Laboratory, April (1976).
36. Darnall, R. R., A. C. Lloyd, A. C. Winer, and J. N. Pitts, Jr., ''Reactivity Scale for Hydrocarbons Based on Reaction with Hydroxyl Radical,'' *Environ. Sci. Technol.* 12:317 (1978).
37. ''Decision Making for Regulating the Chemicals in the Environment,'' National Academy of Sciences, Washington, D.C. (1975).
38. Derwent, R. G., ''On the Comparison of Global, Hemispheric, One-Dimensional and Two-Dimensional Model Formulation of Halocarbons Oxidation by OH Radicals in the Troposphere,'' *Atmospheric-Environment,* 16:551–61 (1982).
39. Donigian, A. S. and M. H. Crawford, ''Modeling Nonpoint Pollution from the Land Surface,'' Hydrocomp. Inc., E.P.A. 600/3–76–083, July (1976).
40. Draggan, S., ''The Microcosm as a Tool for Estimation of Environmental Transport of Toxic Materials,'' *Int. J. Environmental Studies* 10:65 (1976a).
41. Draggan, S., ''The Role of Microcosms in Ecological Research,'' *Int. J. Environ. Stud.* 10:1 (1976b).
42. Dunlap, W. J., ''Isolation and Identification of Organic Pollutants in Groundwater,'' in *Identification and Analysis of Organic Pollutants in Water,* Kieth, H. L. (ed.), Ann Arbor Sci Pub., Ann Arbor, MI (1976).
43. Farmer, W. J., ''Leaching, Diffusion and Sorption,'' in *A Literature Survey of Benchmark Pesticides,* George Washington University Medical Center, Washington, D.C., pp. 185–245 (1976).
44. Farmer, W. J., M. Yang, J. Letey, and W. F. Spencer, ''Residual Management by Land Disposal,'' EPA–600/9/76–015 (1976).
45. Farmer, W. J., K. Igue, W. F. Spencer, and J. P. Martin, ''Volatility of Organochlorine Insecticides from Soil, I: Effects of Concentration, Temperature, Air Flow Rate, and Vapor Pressure,'' *Soil Science Society of America Proceedings* 36:433 (1972).
46. Fay, J. A., ''Physical Processes in the Spread of Oil on a Water Surface,'' Proc. Joint Conf. on Prevention and Control of Oil Spills, API, Washington, D.C. (1971).
47. Finlayson, B. J. and J. M. Pitts, Jr., ''Photochemistry of the Polluted Troposphere,'' *Science* 192:111 (1976).
48. Freed, V. H., C. T. Chiou, and R. Haque, ''Chemodynamics: Transport and Behavior of Chemicals in the Environment—A Problem in Environmental Health,'' *Environmental Health Perspective* 20:55 (1977).
49. Friedlander, S. K., ''Chemical Element Balances and Identification of Air Pollution Sources,'' *Environ. Sci. Technol.* 7:235 (1973).
50. Friedlander, S. K. and H. R. Pruppacher (eds.), ''Report of the Workshop on Research Needs in Intermedia Transport Processes,'' National Center for Intermedia Transport Research, University of California, Los Angeles, April 1–3 (1981).
51. Fuller, B. B., ''Air Pollution Assessment of Trichloroethylene,'' EPA Report No. MTR–7142, PB–256–730, reproduced by the National Technical Information, U.S. Dept. of Commerce, Springfield, VA (1976).
52. Gard, T. C., ''Persistence in Model Ecosystems,'' EPA–600/53–82–030, U.S.

Environmental Protection Agency, Environmental Research Laboratory, Athens, GA (1982).

53. Geophysics Research Forum, ''Studies in Geophysics: Groundwater Contamination,'' National Academy Press, Washington, D.C. (1984).
54. Gillett, J. W., J. Hill, A. Jarrinen, and W. Schoor, ''A Conceptual Model for the Movement of Pesticides through the Environment,'' E.P.A.–600/3–74–024, U.S. Environmental Protection Agency Ecological Research Series (1974).
55. Glotfelty, D. E. and J. H. Caro, ''Introduction Transport and Fate of Persistent Pesticides in the Atmosphere'', in *Removal of Trace Contaminants from the Air,* V. R. Dietz (ed.), American Chemical Symposium Series, ACS, Washington, D.C. (1975).
56. Gillett, J. W. and J. D. Gile, ''Pesticide Fate in Terrestrial Laboratory Ecosystems,'' *Int. J. Environmental Studies* 10:15 (1976).
57. Govers, H., C. Ruepert and H. Aiking, ''Quantitative Structure-Activity Relationships for Polycyclic Aromatic Hydrocarbons: Correlation Between Molecular Connectivity, Physico-chemical Properties, Bioconcentrations, and Toxicity in Daphnia Pulex,'' *Chemosphere* 13:227–36 (1984).
58. Hamaker, J. W., ''Diffusion and Volatilization'', in *Organic Chemicals in the Soil Environment,* Marcel Dekker, New York (1972).
59. Hamaker, J. W. and J. M. Thompson, ''Adsorption'' in *Organic Chemicals in the Soil Environment,* Marcel Goring and J. W. Hamaker (eds.), vol. 1, Marcel Dekker, New York (1972).
60. Hamaker, J. W., ''The Interpretation of Soil Leaching Experiments'' in *Experimental Dynamics of Pesticides,* R. Hague and V. H. Freed (eds.), Plenum Press, New York (1975).
61. Hampson, R. F. and D. Garvin, ''Reaction Rate and Photochemical Data for Atmospheric Chemistry,'' NBS special publication, 513 (1977).
62. Hanna, S. R., ''A Simple Disperison Model for the Analysis fo Chemically Reactive Pollutants,'' *Atmos. Environ.* 7:803 (1973).
63. Hantzicker, J. J., S. K. Friedlander, and C. I. Davidson, ''Material Balance for Automobile Emitted Lead in the Los Angeles Basin,'' *Environ. Sci. Technol.* 9:448 (1975).
64. Harrison, H. L., O. L. Loucks, J. W. Mitchell, D. F. Parkhurst, C. R. Tracy, D. G. Watts, and Y. J. Yannacone, Jr., ''Systems Studies of DDT Transport,'' *Science* 170:503 (1970).
65. Hedden, K. F. and L. A. Mulkey, ''Application of Multimedia Exposure Assessment to Drinking Water'' in *Proceedings of the International Workshop on Exposure Monitoring,* U.S. Environmental Protection Agency, Environmental Monitoring Systems Laboratory, Las Vegas, October 19–22 (1981).
66. Heicklen, J., *Atmospheric Chemistry,* Academic Press, New York (1976).
67. Herbes, S. E. and L. R. Schwall, ''Microbial Transformation of Polycyclic Aromatic Hydrocarbons in Pristine and Petroleum Contaminated Sediments,'' *Appl. Environ. Microbial.* 35:306–16 (1978).
68. Hill, J., IV., and R. G. Weigert, ''Microcosm in Ecological Modeling,'' in DOE Symp. Series #52, J. P. Giesy (ed.), pp. 138–63 (1980).
69. Holm, H. W., H. P. Kollig, L. M. Proctor, and W. R. Payne, Jr., ''Laboratory Ecosystems for Studying Chemical Fate: An Evaluation Using Methyl Parathion,''

EPA–600/53–82–020, U.S. Environmental Protection Agency, Environmental Research Laboratory, Athens, GA (1982).

70. Hushon, J. M., A. W. Klein, W. J. M. Strachan, and F. Schmidt-Bleek, "Use of OECD Premarket Data in Environmental Exposure Analysis for New Chemicals," *Chemosphere* 12:887–910 (1983).
71. Johanson, R. C., J. C. Imhoff, and H. H. Davis, Jr., "User Manual for Hydrological Simulation Program-Fortran (HSPF)," Env. Res. Lab., Athens, EPA 600/9–80–015 (1980).
72. Johanson, R. C., "A New Mathematical Modeling System," in *Fate of Chemicals in the Environment,* R. L. Swann and A. Eschenroeder (eds.), ACS Symposium Series, No. 225, American Chemical Society, Washington, D.C. (1983).
73. Karickhoff, S. W., D. S. Brown, and T. A. Scott, "Sorption of Hydrophobic Pollutants on Natural Sediments," *Water Research* 13:241 (1979).
74. Karickhoff, S. W., "Sorption Kinetics of Hydrophobic Pollutants in Natural Sediments", in *Contaminants and Sediments,* R. A. Baker (ed.), vol. 2, pp. 193–205 (1980).
75. Karickhoff, S. W., "Semi-empirical Estimation of Sorption of Hydrophobic Pollutants on Natural Sediments and Soils," *Chemosphere* 10:833–46 (1981).
76. Kenaga, G. E. and C. A. I. Goring, "Relationship Between Water Solubility, Soil Sorption, Octanol-Water Partitioning and Concentration of Chemicals in Biota," in *Aquatic Toxicology,* ASTM Stp 707, J. G. Eaton, P. R. Parrish, and H. C. Hendrick (eds.), pp. 78–115, American Society for Testing and Materials, Philadelphia, PA, (1980).
77. Khalil, M. A. K., R. A. Rasmussen, and S. D. Hoyt, "Atmospheric Chloroform ($CHCl_3$): Ocean-air Exchange and Global Mass Balance, *Tellus* 35B, 266–74 (1983).
78. Knisel, W. G., "CREAMS: A Field Scale Model for Chemicals, Runoff and Erosion from Agricultural Management Systems," U.S. Department of Agriculture, Washington, D.C. (1980).
79. Kohnke, H., *Soil Physics,* McGraw-Hill, New York (1968).
80. Lee, G. F., "Factors Affecting the Transfer of Materials Between and Sediments," University of Wisconsin, Eutrophication Information Program, Literature Review No. 1 (1979).
81. Lehmann, E. J., "Sediment Water Interaction and its Effect Upon Water Quality, A Bibliography with Abstracts," NTIS/PS–77/0021/4WP, National Technical Information Center, Springfield, VA (1977).
82. Lerman, A. and T. A. Ligtzke, "Uptake and Migration of Tracers in Lake Sediments," *Limnol. Oceanogr.* 20:497–509 (1975).
83. Levy, H., II, "Photochemistry of the Troposphere," *Adv. Photochem.* 9:369–524 (1974).
84. Litchtenstein, E. P., T. T. Liang, and T. W. Fugremann, "A Compartmentalized Microcosm for Studying the Fate of Chemicals in the Environment," *J. Agric. Food Chem.* 26:948 (1978).
85. Lyman, W. J., W. F. Reehl, and D. H. Rosenblatt, *Handbook of Chemical Property Estimation Methods: Environmental Behavior of Organic Compounds,* McGraw-Hill (1982).
86. Mabey, W. R. and T. Mill, "Critical Review of Hydrolysis of Organic Compounds in Water Under Environmental Conditions," *J. Phys. Chem Ref. Data* 7:383 (1978).

87. MacIntyre, F., "The Top Millimeter of the Ocean," *Scientific American* 62–77, May (1974).

88. Mackay, D. and A. T. K. Yeun, "Mass Transfer Coefficient Correlations for Volatilization of Organic Solutes in Water," *Environ. Sci. Technol.* 17:211–17 (1983).

89. Mackay, D., W. Y. Shiu, and R. P. Sutherland, "Determination of Air-Water Henry's Law Constants for Hydrophobic Pollutants," *Environ. Sci. Technol.* 13:333 (1979).

90. Mackay, D., S. Paterson, and M. Joy, "Applications of Fugacity Models to the Estimation of Chemical Distribution and Persistence in the Environment," in *Fate of Chemicals in the Environment,* R. L. Swann, and A. Eshenroeder (eds.), ACS Symposium Series, No. 225, ACS Washington, D.C. (1982).

91. Mackay, D. M., P. V. Roberts, and J. H. Cherry, "Transport of Organic Contaminants in Groundwater," *Environ. Sci. Technol.* 19:384 (1985).

92. Mann, K. H. and R. B. Clark, "Long-Term Effects of Oil Spills on Marine Intertidal Communities," *J. Fish. Res. Board Canada* 35:791 (1978).

93. McClure, V. E., "Transport of Heavy Chlorinated Hydrocarbons in the Atmosphere," *Environ. Sci. Technol.* 10:1,223 (1976).

94. McEwan, M. J. and L. F. Phillips, *Chemistry of the Atmosphere,* John Wiley and Sons, New York (1975).

95. Metcalf and Eddy, Inc., *Wastewater Engineering,* McGraw-Hill, New York (1979).

96. Metcalf, R. L., "Model Ecosystem Studies of Bioconcentration and Biodegradation of Pesticides," in *Pesticides in Aquatic Environments,* M. A. Q. Khan, (ed.), pp. 127–44, Plenum Press, New York (1977a).

97. Metcalf, R. L., "Biological Fate and Transformation of Pollutants in Water," in *Fate of Pollutants in the Air and Water Environments,* I. H. Suffet, (ed.), Part 2, John Wiley and Sons, New York (1977b).

98. Miller, G. C. and R. G. Zepp, "Effects of Suspended Sediments on Photolysis Rates of Dissolved Pollutants," *Water Res.* 13:453–59 (1979).

99. Nash, R. G., "Plant Uptake of Insecticides, Fungicides and Fumigants from Soils," in *Pesticides in Soil and Water,* W. D. Guenzi (ed.), pp. 257–313, Soil Science Society of America, Inc., Madison, WI (1974).

100. National Academy of Science, "Assessing Potential Ocean Pollutants," Washington, D.C. (1975).

101. Neely, W. B., "Estimating Rate Constants of the Uptake and Clearance of Chemicals by Fish," *Environ. Sci. Technol.* 13:1,506–510 (1979).

102. Neely, W. B., D. R. Branson, and G. E. Blau, "Partition Coefficients to Measure Bio-Concentration Potential of Organic Chemicals in Fish," *Environ. Sci Technol.* 8:1,113 (1974).

103. Neely, W. B., "A Material Balance Study of Polychlorinated Biphenyls in Lake Michigan," *Science of the Total Environment* 1:117 (1977a).

104. Neely, W. B., "Material Balance Analysis of Trichlorofluoromethane and Carbon Tetrachloride in the Atmosphere," *Sci. Total Environ.* 8:267 (1977b).

105. Neely, W. B., *Chemicals in the Environment,* Marcel Dekker, New York (1980).

106. Nieuwstadt, F. T. M. and H. Van Dop (eds.), *Atmospheric Turbulence and Air Pollution Modeling,* D. Reidel, Boston (1982).

107. Novotny, V. and G. Chesters, *Handbook of Nonpoint Pollution,* Van Nostrand Reinhold, New York (1981).

108. Oeschger, H., U. Siegenthaler, U. Schotterer, and A. Gugelmann, "A Box Diffusion Model to Study the Carbon Dioxide Exchange in Nature," *Tellus* (2), 168–92 (1975).
109. O'Leary, D. T., K. M. Richter, P. A. Hillis, P. H. Wood, and S. E. Campbell, "Methodology for Estimating Environmental Loadings from Manufacture of Synthetic Organic Chemicals," EPA–600/3–83–064, U.S. Environmental Protection Agency, Environmental Research Laboratory, Athens, GA (1983).
110. Onishi, Y. and S. E. Wise, "Mathematical Model, SERATRA for Sediment Contaminant Transport in Rivers and its Application to Pesticide Transport in Four Mile and Wolf Creeks in Iowa," Batelle Pacific MW Labs, Richland, WA (1979).
111. Parker, B. and G. Barsom, "Biological and Chemical Significance of Surface Microlayers in Aquatic Ecosystems," *Bio. Science,* Jan. 15, pp. 87–93 (1970).
112. Park, R. A., C. I. Conolly, F. R. Albanese, L. S. Clesceri, G. W. Heitzman, H. H. Herbrandson, B. H. Indyke, J. R. Loehe, S. Ross, D. D. Sharma, and W. W. Shuster, "Modeling the Fate of Toxic Organic Materials in Aquatic Environments," EPA–600/53–82–028, U.S. Environmental Protection Agency, Environmental Research Laboratory, Athens, GA (1982).
113. Pasquill, F. and F. B. Smith, *Atmospheric Diffusion,* 3rd ed., Ellis Horwood Ltd., West Sussex, England (1983).
114. Patterson, M. R., T. J. Sworski, A. L. Sjoreen, M. G. Browman, C. C. Coutant, D. M. Hetrick, E. D. Murphy, and R. J. Raridon, "A User's Manual for UTM-TOX: A Unified Transport Model," draft report prepared for the Office of Toxic Substances, US/EPA (1982).
115. Pinder, F. G., "Groundwater Contaminant Transport Modeling," *Env. Sci and Technol.* (1984).
116. Pitts, J. N., Jr., "Mechanisms, Models and Myths: Fiction and Fact in Tropospheric Chemistry," NASA Reference Publication, 1,022 (1978).
117. Pritchard, P. H., "Model Ecosystems," in *Environmental Risk Analysis for Chemicals,* R. A. Conway (ed.), Van Nostrand Reinhold, New York (1982).
118. "Proceedings of the International Workshop on Exposure Monitoring," U.S. Environmental Protection Agency, Environmental Monitoring Systems Laboratory-Las Vegas, published by University of Nevada-Las Vegas, October 19–22 (1981).
119. "Proceedings of the Workshop on Transport and Fate of Toxic Chemicals in the Environment," Norfolk, Virginia, December 17–20, 1978. Office of Environmental Processes and Effects Research, Office of Research and Development, U.S. Environmental Protection Agency, Washington, D.C.
120. Pruppacher, H. R., R. G. Semonim, and W. G. N. Slinn (eds.), *Precipitation Scavenging, Dry Deposition, and Resuspension,* vol. 1; *Precipitation Scavenging,* vol. 2; *Dry Deposition and Resuspension,* Elseville, New York (1982).
121. Pye, V. I. and P. J. Quarles, *Groundwater Contamination in the United States,* University of Pennsylvania Press, PA (1983).
122. Robinson, A. R., "Eddies and Ocean Circulation," *Oceanus* 19 (1976).
123. Seinfeld, J. H., *Air Pollution, Physical and Chemical Fundamentals,* McGraw-Hill, New York (1975).
124. Sekulic, T. S., R. C. Tai, K. R. Boldt, and B. T. Delaney, "Air Quality Effects of A Hazardous Waste Incineration System," *Environmental Progress* 2:47–50 (1983).
125. Shen, T. T. and G. H. Swell, "Air Pollution Problems at Uncontrolled Hazardous

Waste Sites,'' *Civil Engineering for Practicing and Design Engineer* 3:241–59 (1984).

126. Shonh, S. L., ''Mathematical Modeling for Prediction of Chemical Fate'' in *Environmental Risk Analysis for Chemicals,* R. A. Conway (ed.), Van Nostrand Reinhold, New York (1982).
127. Singh, H. B., L. J. Salas, H. Shigeishi, and E. Scribner, ''Atmospheric Hydrocarbons, Hydrocarbons and Sulfur Hexafluoride: Global Distributions, Sources and Sinks'', *Science* 203:899 (1979).
128. Slinn W. G. N., L. Hasse, B. B. Hicks, A. W. Hogan, D. Lal, P. S. Liss, K. O. Munnich, G. A. Sehmel, and O. Vittori, ''Some Aspects of the Transfer of Atmospheric Trace Constituents Past the Air-Sea Interface,'' *Atmospheric Environment* 12:2,055–87 (1978).
129. Smith, J. H., W. R. Mabey, M. Bohonas, B. R. Holt, S. S. Lee, T. W. Chou, P. C. Bomberger, and T. Mill, ''Environmental Pathways of Selected Chemicals in Freshwater Systems'' Parts I and II, E.P.A. 60017–78–074 (1978).
130. Spencer, W. F. and M. M. Claith, ''Vaporization of Chemicals,'' in *Environmental Dynamics of Pesticides,* R. Hague and V. H. Freed (eds.), Plenum Press, New York (1975).
131. Start, G. E. and L. L. Wendell, ''Regional Effluent Dispersion Calculations Considering Spatial and Temporal Meteorolgical Variations,'' NOAA Technical Memorandum, ERL APRL–44, Air Resources Laboratories, 1974.
132. Sugiura, K., S. Sato, and M. Gato, ''Toxicity Assessment Using an Aquatic Microcosm,'' *Chemosphere* 2:113 (1976).
133. Su, C. and Goldberg, E. D., ''Environmental Concentrations and Fluxes of Hydrocarbons,'' in *Strategies for Marine Pollution Monitoring,* Goldberg, E. D. (ed.), pp. 353–74, John Wiley and Sons, New York 1976.
134. Swann, R. L. and A. Eshenroeder, ''Fate of Chemicals in the Environment'', ACS Symposium Series, No. 225, American Chemical Society, Washington, D.C. (1983).
135. Thibodeaux, L. J., ''Mechanisms and Idealized Dissolution Modes for High Density Immiscible Chemicals Spilled in Flowing Aqueous Environments,'' *AIChE, J* 23:544–53 (1977).
136. Thibodeaux, L. J., D. G. Park, and H. H. Heck, ''Chemical Emissions from Surface Impoundments,'' *Environmental Progress* 3:73 (1984).
137. Thibodeaux, L. J., *Chemodynamics,* John Wiley and Sons, New York (1979).
138. Tans, P. P., ''On Calculating the Transfer of Carbon-13 in Reservoir Models of the Carbon Cycle,'' *Tellus* 32, 464–69 (1980).
139. Thompson, A. M., ''Wet and Dry Removal of Tropospheric Formaldehyde at Coastal Site,'' *Tellus* 32, 376–83 (1980).
140. Travis, J. R., ''A Model for Predicting the Redistribution of Particulate Contaminants from Soil Surfaces,'' UŠERDA LA-6035-MS, Los Alamos Scientific Laboratory (1975).
141. Tucker, W. A., A. G. Eschenroeder, and G. C. Magil, ''Air, Land, Water Analysis System: A Multimedia Model for Assessing the Effects of Airborne Toxic Substances on Surface Quality,'' first draft report, prepared by Arthur D. Little, Inc. for Athens Environmental Research Laboratory, US, EPA (1982).
142. Turner, D. B., ''Workbook of Atmospheric Dispersion Estimates'' Office of Air

Programs, pub. AP-26, U.S. Environmental Protection Agency, Washington, D.C., revised (1970).

143. United States, Environmental Protection Agency, "TSCA Candidates List of Chemical Substances," Office of Toxic Substances, Washington, D.C. (1977).

144. Veith, G. D., D. L. Defoe, and B. V. Bergstedt, "Measuring and Estimating the Bio-concentration Factors of Chemicals in Fish," *J. Fish. Res. Board. Can.* 36:1,040 (1979).

145. Vilker, V., personal communication, Department of Chemical Engineering, University of California, Los Angeles (1984).

146. Wallace, A. and W. L. Berry, in "Report of the Workshop on Research Needs in Intermedia Transport Processes", S. K. Friedlander and H. R. Pruppacher (eds.), National Center for Intermedia Transport Research, University of California, Los Angeles, April 1–3 (1981).

147. Ward, D. M. and T. D. Brock, "Environmental Factors Influencing the Rate of Hydrocarbon Oxidation in Temperate Lakes," *Appl. Environ. Microbial* 31, 764–72 (1976).

148. Wiersma, G. B., "Kinetic and Exposure Commitment Analyses of Lead Behavior in a Biosphere Reserve," Technical Report, Monitoring and Assessment Research Center, Chelsea College, University of London (1979).

149. Woodwell, G., P. Craig, and H. Johnson, "DDT in the Biosphere: Where Does It Go?" *Science* 174: 1,101 (1971).

150. Wright, R. T. and J. E. Hobbie, "The Uptake of Organic Solutes in Water," *Limnol. Oceangr.* 10:22 (1965).

151. Zepp, R. G. and D. M. Cline, "Rates of Direct Photolysis in the Aquatic Environment," *Environ. Sci. Technol.* 11, 359–66 (1977).

INDEX

ABOUT THE EDITORS

Sidney Draggan, Ph.D., is an ecologist with special interest in toxic chemical testing and control. He has served as research ecologist at Oak Ridge National Laboratory, senior visiting research fellow at the Monitoring Assessment Research Center in London, U.K., ecological effects team leader with the U.S. Environmental Protection Agency, policy analyst with the National Science Foundation, division of policy research and analysis, and is now associate program manager for polar biology and medicine at the National Science Foundation.

John J. Cohrssen is attorney advisor, the Council on Environmental Quality, Executive of the President, Washington, DC, and was responsible for the *Report on Long-Term Environmental Research and Development*. Mr. Cohrssen specializes in regulatory law with emphasis in health and environmental areas. He is the author of numerous professional publications, and a member of federal, state, and local bar associations. Mr. Cohrssen holds the following degrees: B.S. City College of New York, 1961, M.Sc. McGill University, 1963, J.D. George Washington University, 1967.

Richard E. Morrison is a senior economist and science policy analyst with the National Science Foundation (NSF), where he is responsible for the preparation of science policy analyses and working papers of interest to the National Science Board Chairman, the NSF Director, and other NSF officials. His expertise is the interaction of federal laws, regulations, and policies with industrial development and with commercialization of new technological products and services. He also assists in the development of short- and long-range policy alternatives and strategies.